MW00965747

A NOBEL AFFAIR

EDITED AND TRANSLATED
BY ERIKA RUMMEL

A Nobel Affair

The Correspondence between
Alfred Nobel and Sofie Hess

UNIVERSITY OF TORONTO PRESS
Toronto Buffalo London

© University of Toronto Press 2017
Toronto Buffalo London
www.utppublishing.com
Printed in the U.S.A.

ISBN 978-1-4875-0177-8

Printed on acid-free, 100% post-consumer recycled paper
with vegetable-based inks.

Library and Archives Canada Cataloguing in Publication

Nobel, Alfred Bernhard, 1833–1896
[Correspondence. Selections. English]
A Nobel affair : the correspondence between Alfred Nobel and Sofie Hess /
edited and translated by Erika Rummel.

Includes bibliographical references and index.
ISBN 978-1-4875-0177-8 (hardcover)

1. Nobel, Alfred Bernhard, 1833–1896 – Correspondence.   2. Hess, Sofie –
Correspondence.   3. Chemical engineers – Sweden – Correspondence.
I. Rummel, Erika, 1942–, editor, translator   II. Hess, Sofie – Correspondence.
Selections. English   III. Title.

TP268.5.N7A4 2017   660.092   C2017-901069-7

University of Toronto Press acknowledges the financial assistance
to its publishing program of the Canada Council for the Arts and
the Ontario Arts Council, an agency of the Government of Ontario.

Canada Council   Conseil des Arts
for the Arts   du Canada

ONTARIO ARTS COUNCIL
CONSEIL DES ARTS DE L'ONTARIO
an Ontario government agency
un organisme du gouvernement de l'Ontario

Funded by the   Financé par le
Government   gouvernement
of Canada   du Canada

Canadä

Contents

Acknowledgments

I wish to express my gratitude to the readers of University of Toronto Press for their close reading and valuable suggestions and to Richard Ratzlaff for his encouragement. I would also like to thank Lena Anime in the National Archive in Stockholm and Christina Sandahl in the National Library of Sweden for their assistance in supplying me with pertinent materials.

A NOBEL AFFAIR

Introduction

The letters exchanged between Alfred Nobel and Sofie Hess, his Viennese mistress, span almost two decades, from 1877 to 1896. The couple wrote to each other in German.[1] Surprisingly, given Nobel's prominence, this is the first English translation of their correspondence in its entirety – 221 letters by Nobel and 41 letters by Hess.[2]

After her lover's death, Hess offered to sell the collection to the Nobel Foundation, perhaps because she was dissatisfied with the annuity granted her under the terms of Nobel's will.[3] To prevent publication of the correspondence, which does not always reflect well on Nobel, Ragnar Sohlman, one of the executors, negotiated a settlement.[4] Sofie

---

1 Nobel demonstrated an excellent grasp of the language; his lover, whose education had been neglected, tended to "mangle" her native language, as he noted in Letter 33. Among her egregious spelling mistakes we find "egsistirt" for "existiert" (SH Letter 1), "aprospos" (repeatedly) for "apropos" (SH Letter 6), and "dasperart" for "desperat" (SH Letter 16).

2 There is a Swedish translation of Alfred Nobel's letters, interspersed with some of Sofie Hess's: *Mitt hjärtebarn: De länge hemlighållna breven mellan Alfred Nobel och hans älskarinna Sofie Hess*, trans. Vilgot Sjöman (Stockholm, 1995). Excerpts from Nobel's and Hess's letters appear in the English translation of Kenne Fant, *Alfred Nobel: A Biography*, trans. M. Ruuth (New York, 2014).

3 She had asked for an annuity of ten thousand florins (SH Letter 34); Nobel left her only six thousand florins in his will. As he pointed out, a family of four could comfortably live on two thousand florins a year (Letter 197). See T. Cvrcek, "Wages, Prices, and Living Standards in the Habsburg Empire 1827–1911," *Journal of Economic History* 73/1 (2013), 1–37.

4 See Ragnar Sohlman, *The Legacy of Alfred Nobel: The Story behind the Nobel Prizes* (London, 1983), p. 78: "[The executors] were anxious to avoid further trouble, especially anything which might reflect badly on Nobel himself – and there was no knowing what scandal might be brought to light by the publication of the letters."

Hess was paid the substantial sum of twelve thousand florins – twice her annuity – and agreed not to comment publicly on her relationship with Nobel. The letters became the property of the Foundation and remained under wraps until 1976, when they were deposited in the Swedish National Archives and became accessible to scholars.[5]

During his lifetime, Nobel's fame rested on his inventions and on his success as a businessman. His correspondence may therefore be regarded as a primary source for historians interested in nineteenth-century economy, science, and technology. The letters Nobel wrote to Sofie Hess do indeed contain references to his scientific work and his business dealings, but they are too vague to contribute significantly to what we already know from other sources. They do, however, offer important new insights to readers interested in the biographical details of his life and to scholars of nineteenth-century social history. The fact that the Nobel Foundation paid Hess for the letters and kept them private for decades is in itself an indication of their historical significance. The Foundation was intent on reinforcing Nobel's image as a high-minded visionary and philanthropist. His correspondence with Hess, by contrast, shows up his all-too-human failings. Reading his letters, we encounter a man obsessed with work, whose tireless pursuit of innovation and business ventures took a heavy toll on his physical and psychological health, a man who was paranoid, cranky, sarcastic, bigoted, and old beyond his years. The discretion exercised by the Foundation led to misinterpretations, such as the suggestion that his relationship with Hess was platonic – untenable on the evidence of their letters – and to a one-sided picture of Nobel's personality.[6]

---

5 My translation is based on the text in the Alfred Nobels Arkiv Ö I-5. The archive contains, in addition to Alfred Nobel's and Sofie Hess's letters (cited as "SH," followed by the number of the letter), twenty-one letters from Sofie's father, Heinrich Hess (cited as "HH," followed by the number of the letter) and assorted letters to Nobel from Sofie's sisters Amalie and Bertha, from her brother-in-law Albert Brunner, and from her husband Nikolaus Kapy. Some of these letters are included in the appendix below.

6 Henrik Schück, for example, completely ignored Nobel's long-standing relationship with Sofie Hess in his biography *The Life of Alfred Nobel* (Uppsala, 1926; English translation London, 1929). Sohlman quotes him as saying that Nobel's mother was his great love: "He appears to have had no other – with the possible exception of a girl mentioned in one of the poems of his youth" (Legacy, p. 52). Similarly, Erik Bergengren in *Alfred Nobel: The Man and His Work* (New York, 1960) refers to Nobel's lover only as "H," as if he wanted to minimize her existence. He likens Nobel's attitude towards Hess as that of "a kindly social reformer of the Victorian school with pointer in hand"

The correspondence sheds light not only on Nobel's character and his relationship with Hess, but more generally on the position of mistresses in nineteenth-century Europe. The lives of women supported by wealthy bourgeois remain a somewhat neglected area of research, lost between the well-documented subjects of prostitution[7] at one end of the scale and the lives of celebrated courtesans at the other.[8] Sofie Hess does not fit either category. She has little in common with Lillie Langtry or Cora Pearl, who like her had migrated to Paris and soon acquired celebrity status.[9] Hess was stylish and vivacious, but no match for the wit, enterprise, and love of spectacle exhibited by the grand *maitresses*. Nor was this what Nobel wanted from her. He was looking for relaxation and was notably averse to public displays. Indeed, he wanted Hess to provide him with a cozy, home-like atmosphere and with quiet cheer and comfort.[10] Although Hess had no independent source of income and fully relied on Nobel to look after her expenses, their relationship cannot be described simply as an exchange of sex for money, such as occurs between a prostitute and her client. Indeed, Nobel continued to support Hess after their physical relationship ended, and Hess in turn was loyal to him for many years until he himself suggested she look for another man and find a suitable marriage partner.

(p. 170). Their relationship, he says, was a "touching example of Nobel's daily avuncular care, instructive and pecuniary" (p. 199). A recent publication of the Nobel Museum (Ulf Larsson, *Alfred Nobel: Networks of Innovation* [Stockholm, 2008]) perpetuates the impression that Nobel was merely searching for friendship and saw in Sofie Hess "someone to be together with, to share tenderness and confidences with" (p. 165).

7 For a study and full bibliography on the subject of prostitution in Paris, where Nobel lived, see Andrew I. Ross, *Urban Desires: Practicing Pleasure in the "City of Light" 1848–1900* (doctoral diss., University of Michigan, 2011, http://deepblue.lib.umich.edu/bitstream/handle/2027.42/89671/aiross_1.pdf?sequence=1).

8 See, for example, Elizabeth Abbott, *Mistresses: A History of the Other Woman* (London, 2010); Victoria Griffin, *The Mistress: Histories, Myths, and Interpretations of the "Other Woman"* (New York, 1999), and, more generally, Priscilla Robertson, *An Experience of Women. Pattern and Change in Nineteenth-century Europe* (Philadelphia, 1982).

9 See Katie Hickman, *Courtesans: Money, Sex, and Fame in the 19th Century* (New York, 2003).

10 See Letter 130 in which he praises compassion and cheer. In Letter 78 he notes that "the true woman" adapts to a man's feelings; in Letter 30 he states that the role of the woman is to "sweeten" a man's life; in Letter 164 he compliments Hess on the comfortable atmosphere of her apartment and the peace and quiet she provided for him on a visit.

Unlike the lives of the great *maitresses* of Paris, which are generally reflected in the comments of others or described in carefully construed autobiographies referring to events long past, the lives of Hess and her lover are given immediacy through letters written "in real time." It is tempting to compare (or rather, contrast) their correspondence with another collection: the letters exchanged between Victor Hugo and Juliette Drouet, his "Muse" and mistress. Such a comparison is suggested by the fact that the two men, Nobel and Hugo, were on visiting terms and familiar with each other's life and work.[11] Drouet, a well-known model and actress, wrote numerous letters to her lover, pouring out a flood of passionate words. After one romantic evening, she writes to Hugo: "I drank everything you left in your glass ... I shall eat with your knife and sup from your spoon. I kissed the spot where you rested your head ... I surround myself and steep myself in all that was close to you."[12] Sofie Hess, by contrast, had to be prompted to write. Her letters to Nobel focus on mundane subjects – most often his and her poor health and her financial needs. They are devoid of passion. Any expressions of affection and longing are trite and lacking in emotional warmth, but if Hess was no "Muse," Nobel was no Hugo. His letters were hardly of a kind to evoke the sort of passion Hugo called forth with his "battering-ram of adjectives" describing his love for Drouet as "insane, absurd, extravagant, malicious, jealous, nervous – whatever you will – but it is love ... I long to be lying at your feet, kissing them."[13]

Like Nobel, Hugo tried to hide his lover away and isolate her in a rented apartment to keep the affair out of the public eye (and in Hugo's case, out of his wife's eye). How scandalous was it for a man to have a mistress in nineteenth-century Paris? It seems the position of the kept woman was ambiguous. Alfred Delvau, a contemporary of Nobel and Hugo and the author of *Les Plaisirs de Paris*, remarks that men were proud of their conquests. "Going to the Bois was an excellent occasion to show off their horses or their mistress" – but only if the woman of their choice was "*à la* mode," that is, was chic and had a "laughing grace of manner."[14] Nobel may have had such expectations of Sofie Hess and found that she could not fulfil them. But whether or not a

---

11 See for example Letters 52 and 60. Nobel had fifteen works of Hugo in his private library. See http://www.nobelprize.org/alfred_nobel/library/fiction-fr.html.
12 Quoted in Graham Robb, *Victor Hugo: A Biography* (New York, 1997), p. 247.
13 Robb, *Victor Hugo*, p. 188.
14 Alfred Delvau, *Les Plaisirs des Paris: Guide pratique et illustré* (Paris, 1867), pp. 28, 285.

man was proud of his mistress, a certain protocol had to be followed to avoid scandal. Listing fashionable restaurants in Paris, Delvau distinguishes between those "where you would want to take your wife and others where you would take your mistress."[15] Proprieties had to be observed. It was for this reason that Nobel, even though he was unmarried, kept a separate establishment for Sofie Hess. When they travelled, however, they shared accommodations, and a mock-propriety was observed by referring to Hess as "Mrs. Nobel."[16] Far from causing scandal, their relationship was tolerated, and Hess commanded a certain measure of interest and respect as the companion of an important man. Similarly, Hugo's mistress Juliette Drouet, who died of cancer in 1883, was honoured posthumously by politicians, literary men, and indeed Nobel himself, whose name appears in the condolence book.[17] Thus Sofie Hess was by no means ostracized because she was the mistress of Nobel. It was only in later years, when she lost his protection and was burdened with an illegitimate child fathered by another man, that she was discriminated against and suffered the fate of a "fallen" woman.[18]

Nobel made Sofie Hess's acquaintance relatively late in life, at the age of forty-three. He was born in Stockholm in 1833, the fourth son of entrepreneur and inventor Immanuel Nobel and his wife Carolina Andrietta. He and his siblings grew up in poverty until the elder Nobel emigrated to St. Petersburg and prospered there as a manufacturer of military and construction equipment. Alfred therefore received an excellent private education. He took a special interest in chemistry[19] and language studies, acquiring fluency in English, French, Russian, and German.[20] At the age of eighteen he embarked on educational travels in Western Europe and America. In Paris, he worked under the direction of the chemist Théophile Jules Pelouze and in New York under John Ericsson, who built the first armoured warship. In 1859 Nobel returned to his native Sweden with his father, who had suffered financial setbacks, and built an explosives factory near Stockholm. Over the next

---

15 Delvau, *Les Plaisirs des Paris*, p. 113.
16 See Letters 32, 185, 186.
17 Robb, *Victor Hugo*, p. 318.
18 SH 17.
19 His tutor was Nikolaj Sinin (1812–1880), professor of chemistry at the university of St. Petersburg.
20 He also tried his hand at literary productions. In 1851 he wrote a long autobiographical poem in English, entitled *The Riddle*. In the last year of his life he wrote a play in Swedish, which was printed (*Nemesis*, Paris, 1896) but never staged.

thirty years, he obtained more than three hundred patents, among them one for "blasting oil" in 1863.

Nobel's brothers, Ludvig and Robert, remained in Russia and continued with the manufacture of military equipment. In 1873 they acquired oil fields near Baku and in 1879 they founded Branobel, an oil company, in which all three brothers held shares.[21] Nobel himself formed a network of corporations and factories for the production of explosives in Sweden, Germany, Austria, Scotland, France, Italy, and America. For many years (1873–1891) he made his home in Paris at 53 Avenue Malakoff and conducted his tests in a laboratory in nearby Sévran. The manufacturing process was marred by accidents at first[22] and motivated Nobel to develop a safer product by mixing nitroglycerine with kieselgur (patented in 1867 as "dynamite"). His wide-flung business empire required Nobel to travel incessantly and took a toll on his already compromised health. Victor Hugo quipped that he was the wealthiest vagabond in Europe.[23] After the reckless speculations of Paul Barbe, Nobel's French partner, involved Nobel in a financial scandal and exposed him to sharp attacks in the press,[24] he moved to San Remo on the Italian Riviera, where he died in 1896.

Designed originally for industrial blasting operations, dynamite was more commonly employed in warfare, earning Nobel the sobriquet "merchant of death,"[25] but he was also hailed as a man of genius and a philanthropist. In accordance with the directions in his will, the bulk of his estate was used to fund annual prizes for advancements in physics, chemistry, medicine, literature, and – ironically – the promotion of peace. The peace prize may have been inspired by Bertha von Suttner, who was briefly Nobel's secretary but is better known today as an activist for pacifism. She carried on a correspondence with Nobel from 1883 to his death and was the first woman to receive the Nobel Prize for Peace in 1905.[26]

---

21 For the history of Branobel see Robert W. Tolf, *The Russian Rockefellers: The Saga of the Nobel Family and the Russian Oil Industry* (Stanford, 1976).
22 His youngest brother, Emil, was killed at the age of twenty-one in an explosion at the Nobel factory in Heleneborg near Stockholm.
23 Quoted in Bergengren, *Alfred Nobel*, p. 160.
24 See Letters 148, 165, 171.
25 Bergengren, *Alfred Nobel*, p. 161.
26 For their correspondence see Edelgard Biedermann, *Der Briefwechsel zwischen Alfred Nobel and Bertha von Suttner* (Hildesheim, 2001).

Nobel often complained that he was lonely and friendless, and in later years tried to blame Sofie Hess for his isolation: "You have so tarnished my name that I, who do nothing but work and help others, am discredited and must live an isolated life."[27] It seems, however, that he was solitary by nature. Bertha Suttner described him as a man "who did not receive many visitors and rarely went out into the world. He was a hard worker but somewhat shy of people. He hated all small talk."[28] Similarly, Sofie Hess countered his complaint about being left alone with the statement that he "was able to live alone and be content and happy."[29] In fact, Nobel was frequently unhappy and constitutionally depressed – plagued by the "spirits of Niflheim," as he termed his condition in a letter to his brother Ludvig.[30] Nobel had great affection for his mother, but apart from her, Sofie Hess was the only woman who could claim to have a close personal relationship with him.

Sofie Hess (1851–1919) was of Jewish parentage. She was the oldest child of Heinrich Hess, a dealer in timber from Celje (Slovenia) and his wife Amalie. Nobel met Sofie Hess in 1876 in Baden (near Vienna), where she worked in a flower shop.[31] She followed Nobel to Paris and lived first at 1 Rue Newton and, from 1880 on, at 10 Avenue d'Eylau, in apartments Nobel rented for her.[32] Sofie Hess was twenty-six when she made Nobel's acquaintance, but he was under the impression that she was much younger and made her sign a note indicating that he had counselled her to return to her parents: "I acknowledge that Mr. Nobel has [tried to] persuade me to return to my father in Celje, and that it is not his fault if I don't do it."[33]

For the next fourteen years the pair lived together after a fashion. Sofie Hess called it "living together" at any rate, although she was

---

27 Letter 157; see also Letter 111.
28 Suttner, "Erinnerungen an Alfred Nobel," published in *Neue Freie Presse,* 12 January 1897, p. 1.
29 See Letter 157.
30 Quoted in Fant, *Alfred Nobel,* p. 197. In Norse mythology "Niflheim" is the land of mist.
31 By that time her mother was dead and her father had remarried. According to Heinrich Hess, Sofie left home because of strained relations with her stepmother. See Appendix, p. 275.
32 The lease of 1880 shows that the apartment in the Rue d'Eylau consisted of an antechamber, a dining room, a large and a small salon, three bedrooms, a kitchen and office, toilets and bathrooms, and servants' quarters.
33 Dated Vienna, 10 September 1877 (quoted in Sjöman, *Mitt hjärtebarn,* p. 26).

rarely resident in Paris.[34] Nobel disapproved of her constant travelling, her "gallivanting around" fashionable spas all over Europe.[35] He himself joined her at those spas, however, in the vain hope of curing his rheumatism. Almost every letter he wrote contains complaints about his poor health. Apart from rheumatism, he suffered from scurvy, migraines, a nervous stomach, and in the last years of his life from what he termed "heart spasms."[36]

Taking the waters had been an aristocratic pastime at first, but in the nineteenth century the practice was imitated by wealthy bourgeois.[37] Thus John Murray, the author of a *Handbook for Travellers to the Continent*, writes in 1840 that "an excursion to a watering place in the summer is essential to life, and the necessity of such a visit is confined to no one class in particular."[38] Indeed the resorts attracted not only the nobility and the middle class but also the demi-monde[39] and, as Francis Palmer pointedly noted in 1903, "Even the ubiquitous Jew is not wanting ... although [in Carlsbad] they have to take the waters in the early morning, an hour or two before the regular opening for the fashionable visitors."[40]

It is clear from the correspondence that Sofie Hess was combining therapy with amusement. After all, her ailment (anemia?) wasn't entirely physiological. She often complained of low spirits. Nobel encouraged

---

34 SH 20; see also her father's letter to Nobel after their break-up: "At least try to live together with her for a little while longer" (HH 16 in Arkiv ÖI-5, undated).

35 E.g., Letters 66, 72. Hoping perhaps for more permanence, he bought a villa in Ischl, a fashionable watering place in Austria.

36 Letter 127. Sofie's ailments remain undefined during the years she lived with Nobel and travelled to spas with him or on her own. In 1892 she wrote that she suffered from "congestion, vertigo, and anemia" (SH Letter 15); in 1895 she complained of bronchitis (SH Letters 27, 36) and wrote that she had suffered an "embolism" (SH Letter 35). In 1888 her father told Nobel that she was ill with peritonitis (in a letter dated 17 October = HH 9 in Arkiv ÖI-5).

37 On spa culture in nineteenth-century Europe see, for example, Jill Steward, "The Role of Inland Spas as Sites of Transnational Cultural Exchanges, 1750–1870," in *Leisure Cultures in Urban Europe, c. 1700–1870: A Transnational Perspective*, ed. Peter Borsay and J.H. Furnée (Manchester, 2015), 234–59.

38 John Murray, *Handbook for Travellers to the Continent* (London, 1840), p. 217.

39 Jill Steward, "The Spa Towns of the Austro-Hungarian Empire and the Growth of Tourist Culture: 1860–1914," in *New Directions in Urban History: Aspects of European Art, Health, Tourism and Leisure since the Enlightenment*, ed. Günther Hirschfelder et al. (Münster, 2000), 87–126. For the increase in the number of "tainted ladies" at Carlsbad, Marienbad, and Meran, see pp. 110–11.

40 Francis Palmer, *Austro-Hungarian Life in Town and Country* (London, 1903), p. 432.

her to seek out company and enjoy herself.[41] Most spa visitors did just that. They mingled with fashionable society, went on outings, and attended dances and theatrical performances in addition to taking the prescribed treatments. Thus Francis Palmer commented that Carlsbad had turned into a centre of recreational tourism: "It is not supposed to be a pleasure resort. The object of all visitors is, at least ostensibly, the restoration of health that has broken down under the stress of society functions, or political life, overwork or study, or the cares and worries inseparable from the existence of great financiers."[42]

Each spa town had its distinct character. Ischl, where Nobel bought a villa in 1879, was a quiet town. It is described in 1880 as "simple and decorous ... Ischl does not gamble or riot, or conduct herself madly in any way; she is a little old fashioned still in a courtly way; she has a little rusticity still in her elegant manners; she is homely whilst she is so visibly of the *fine fleur*."[43] The Austrian-Hungarian spas, moreover, were tightly regulated. Only licensed physicians were permitted to dispense medication. Nobel, who understood the effects of the chemicals present in the water,[44] perhaps also appreciated the scientific approach of the Austrian practitioners. They considered the quantity and temperature of the water and controlled the time patients spent submerged in baths, while spas in England, for example, allowed unlimited drinking and bathing.[45]

The subject of travel is prominent in Nobel's letters, but it is only one of many aspects of life on which he touches in the correspondence. He refers to his readings and his literary tastes, his social engagements, his domestic arrangements in Paris and his laboratory at Sevran, his business affairs in Germany, Scotland, Belgium, and Russia, and his extended family in Sweden and Russia. Most of all, the letters offer intimate glimpses of his relationship with Hess. They show the trajectory of their liaison, which began as a love affair and ended in a fitful patriarchal relationship.

---

41 Letters 8, 36, 70.
42 Palmer, *Austro-Hungarian Life*, p. 127. See also Douglas P. Mackaman, "The Tactics of Retreat: Spa Vacations and Bourgeois Identity in Nineteenth-Century France," in *Being Elsewhere: Tourism, Consumer Culture, and Identity in Modern Europe and North America*, ed. Shelley Baranowski and Ellen Furlough (Ann Arbor, 2004), 35–62. Spas in the nineteeenth century changed from "outposts of aristocratic sociability and doctoring to become centerpieces of the modern vacation industry" (p. 36).
43 Ouida (M.L. de La Ramé), quoted in Jill Steward, "The Spa Towns," p. 87.
44 See Letters 17, 19, 21.
45 Steward, "The Spa Towns," pp. 91–2.

In the first two years of their acquaintance, Nobel wrote to Sofie almost daily while away on business. His letters are full of endearments and expressions of longing, but already in 1879 we find the first intimations of jealousy, annoyance, and exasperation. Nobel suspected Sofie of an indiscretion with his nephew Emmanuel. The young man (nineteen at the time) called it a misunderstanding, and Sofie's father likewise protested the allegations, but Sofie's sister Amalie remarked that she "betrayed Nobel with every waiter."[46] By 1881 the couple appeared to be fully reconciled. Over the next years, however, there was a gradual shift in their relations. Nobel increasingly tried to make himself out as Sofie's protector and a father figure rather than a lover. Looking back on their relationship in 1887, he claimed that he "adopted [her], so to speak.[47] There can be no doubt, however, that the couple's relationship was sexual to begin with. There are frequent references in Nobel's letters to Sofie's menses, and Sofie herself refers to the difficulty of finding a husband after having been Nobel's mistress (*Mätresse*) for so many years.[48]

Initially Nobel may have had intentions to marry Hess – he made plans to introduce her to his mother[49] – but he soon came to the conclusion that their relationship had no future and made it clear that they were incompatible. He was searching for someone with whom he could share his life, he wrote, but "that someone cannot be a woman ... whose outlook on life and whose intellectual interests have little or nothing in common with mine" (Letter 24). He complained about Sofie's lack of culture and her idleness: "You neither work, nor write, nor read, nor think" (Letter 35). Yet he did not end their relationship. It seems that he was simultaneously repelled and attracted by Sofie's frivolity and thoughtlessness: "That's the nice thing about you – the complete absence of reason" (Letter 94). Playing Pygmalion, he attempted, without much success, to improve Sofie's education and to prompt her to learn

---

46 See Emmanuel Nobel's letters to Sofie, dated 16 and 28 January 1879 (quoted in Sjö-
    man, *Mitt hjärtebarn*, pp. 77–8) and Amalie Brunner's remark (quoted in Sjöman, p. 66).
    Heinrich Hess defends his daughter: Sofie was much afflicted by this suspicion and
    is still willing "to justify herself in the presence of your nephew" (Appendix, p. 275).
47 Letter 115.
48 For references to her menses see Letters 24, 33, 65, 78. She refers to her years as
    Nobel's *Mätresse* at SH 20.
49 Letter 10. He also introduced her to his brothers. See, e.g., Letters 60, 70, 72, 77.

French.[50] In later years he referred contemptuously to her "microscopic brain" (Letter 197) and blamed her for his own intellectual decline. It was the result of associating with her, he wrote: "I have sacrificed [to you] my intellectual life, my reputation which always rests on our association with others, my whole interaction with the cultured world" (Letter 43).

In line with the patriarchal role he assumed, Nobel frequently referred to himself as Sofie's "old uncle."[51] This was a common euphemism to characterize relations between older men and their young lovers, but also seems to reflect a genuine weariness and the realization that he was growing old. Other ways of signing off on his letters, such as "Brummbär" (grouchy-bear) and "Grübler" (ruminator, melancholic), acknowledged in an apologetic manner his morose and gloomy nature.[52]

Nobel rarely used Hess's given name in the first years of their correspondence. He addressed her as his dear "little child" or "little toad" or used diminutives of her name ("Sofferl," "Soffiecherl"). Sofie reciprocated by calling him occasionally "Bubi" (little boy). He in turn called himself her "boy," though usually in a slightly ironic tone.[53]

Around 1884 their relationship began to unravel. He began to complain about Sofie's use of his name[54] – unauthorized, he claimed, although he himself addressed letters to her as "Madame Nobel."[55] He also allowed sarcasm to creep into his letters[56] and made anti-Semitic remarks about the Hess family, now living in Vienna.[57] These aspersions reflect the prevailing climate in Europe. In Paris anti-Semitism had become virulent by the time Edouard Drumont published his racist best-seller *Le France Juive* (1886) and proclaimed his theory that a Jewish conspiracy was afoot: "They want to bring about a violent change in

---

50 See, e.g., Letter 9 in which Nobel corrects Sofie's French. He also alludes to his educational efforts in a letter to Alarik Lindbeck. Expressing himself in unvarnished terms, he writes: "Women are interesting, but even when you are no longer able to stop their hole, you gladly stop their ignorance – the brain stands up longer than the prick" (quoted in Slöman, *Mitt hjärtebarn*, p. 51).

51 E.g., Letters 39, 52.

52 E.g., Letters 11, 20, 16, 24.

53 Hess in SH 4; Nobel in Letters 45, 60, 77.

54 E.g., Letters 177, 181, 184.

55 E.g., Letter 32.

56 Letters 79, 80, 81, 123, 146, 202.

57 E.g., Letters 126, 146, 177, 181, 193.

ideas, manners, and traditional beliefs in this country ... I am but a modest messenger of things to come."[58]

In Vienna, Jews fleeing the Russian pogroms in the 1880s swelled the Jewish population, doubling it between 1870 and 1890. This influx of refugees caused resentment among workers, who saw them as competitors in the job market. Lead articles in rightist papers promulgated a conspiracy theory much like Drumont's, declaring that the Jews would soon be dominating the world. "In a few decades, perhaps, they will become the exclusive lords of our financial markets. The first step to the throne has been taken ... How will we be saved from the claws of the usurers? ... We are looking at complete enslavement."[59] The conspiracy theory is also reflected in the speeches of the politician Georg von Schönerer, the founder of the extremist Pan-German Party and chief representative of anti-Semitism in Austria. He was unapologetic about his racism: "Anti-Semitism should not be seen as a regrettable or shameful symptom, but rather as a pillar of national thought and chiefly as a demand for true ethnic sentiment."[60] He enjoyed wide support among the lower middle class and the student fraternities. In 1881 he led a band of hooligans to ransack the offices of the liberal newspaper *Neues Wiener Tagesblatt*, which he labelled a "shameful Jewish rag."[61] In response to these criminal excesses, liberal thinkers in Vienna formed an association to combat racism: the Verein zur Abwehr des Antisemitismus, founded in 1891.

It should be noted that the founder of the Verein was Arthur Suttner, the husband of Bertha Suttner, who had kept up a correspondence with Nobel and visited him, together with Arthur, in 1887 and 1892. It is not surprising that Nobel refrained from disparaging remarks about Jews in letters to her. Sofie Hess's failure to respond to his anti-Semitic diatribes is surprising, however. Far from protesting against his cutting remarks, she used abusive language against Jews herself and, in 1894,

---

58 Edouard Drumont, *Le France Juive: Histoire Contemporaine* (Paris 1886), p. 85.
59 *Österreichischer Volksfreund*, 19 February 1881, p. 2.
60 Speech quoted in Lisa Kienzl, *Nation, Identität und Antisemitismus: Der deutschsprachige Raum der Donaumonarchie 1866–1914* (Graz, 2014), p. 122.
61 Reacting to Schönerer's provocations, the liberal paper *Neue Freie Presse* labelled his anti-Semitism a political move: "The Jew is merely a means of turning the masses against liberalism" (6 April 1882). For a general account of the position of Jews in Vienna at the time, see Robert Wistrich, *The Jews of Vienna in the Age of Franz Joseph* (New York, 1989).

converted to Protestantism – a move she naively expected to bring her closer to Nobel.[62] Although it may be difficult for us today to understand her attitude, which amounts to self-hatred, it is a well-documented phenomenon among converts and perhaps akin to the Stockholm syndrome. In an effort to be assimilated into the dominant Christian society, Jews suppressed their roots, as the Viennese merchant Sigmund Mayer acknowledged in his memoirs: "I had quite forgotten that I was a Jew until I made this unpleasant discovery, prompted by anti-Semitism."[63] This "forgetfulness" was a first step towards identification with the dominant and often hostile gentile population. Whether assimilation was forced or not, it turned into a "spiral that drove Jews further into self-denial ... and at the end of that path we often find undisguised Jewish self-hatred."[64]

In spite of Hess's odious remarks about fellow Jews, she always remained loyal to her own family and was open-handed in supporting them financially. If this close relationship was an irritant to Nobel, her spending habits were another bone of contention. Nobel called Hess his "great devourer of banknotes."[65] Although it was he who introduced Sofie to a life of luxury, her excesses certainly justified his complaints. She was unable to control her spending and pawned her jewellery or contracted debts whenever she ran short of money. In 1894, she was formally placed under the *Kuratel* (trusteeship) of Julius Heidner, a director of Nobel's company in Vienna, where she lived at the time. Her dependence on Heidner, who doled out her annuity in monthly instalments, frustrated Sofie and led to some remarkably ill-natured and racist comments about Nobel's legal representatives in Vienna.[66]

---

62 SH 20: "so that we are now closer to each other than ever."
63 Sigmund Mayer, *Ein jüdische Kaufmann 1831–1911* (Vienna, 1988), p. 381 (my translation; quoted in the original by Michael Ley, *Abschied von Kakanien: Antisemitismus und Nationalismus im Wiener Fin de siècle* [Vienna 2001], p. 195).
64 Adolf Gaisbauer, *Davidstern und Doppeladler: Zionismus und jüdischer Nationalismus in Österreich 1882–1918* (Cologne, 1988), p. 23, my translation
65 Letter 118. For other complaints about her spending habits see, e.g., Letters 158, 170, 197. In Letter 158 he mentions the figure of 48,267 francs spent by Hess in six months. Compare this, however, with the sums spent by her contemporary, the notorious Cora Pearl: "In just two weeks at Vichy her household expenses topped thirty thousand francs (56,000 pounds)" (Hickman, *Courtesans*, p. 132).
66 An official notice of the trusteeship appeared in the *Central-Anzeiger für Handel und Gewerbe* on 10 July 1894. For Sofie's spiteful remarks on Nobel's legal representatives see, e.g., SH 22, 20 ("that red Jew be damned ... and Philip protects him because he too is a baptized Jew").

From 1882 on Nobel counselled Sofie, first covertly and then openly, to find a husband[67] and seemed to be willing to make financial arrangements to further this goal.[68] The situation has a curious parallel in Dostoevsky's *Idiot* (1868), in which the wealthy aristocrat Totsky supplies his mistress Nastassya Filippovna with a dowry of seventy-five thousand rubles when he wants to end their relationship. His desire to arrange a marriage for his lover no doubt reflects actual practices at the time and is not entirely irrelevant to Nobel's situation, given his ties to Russia and his interest in Russian literature.[69]

Nor is the role that Sofie's father played in this affair without precedent. Heinrich Hess freely discussed Sofie's financial settlement with Nobel. In 1887 he asked for a personal meeting to review the matter and asserted his daughter's right to financial support. Nobel could not simply discard her "after an intimate relationship of ten years," he said.[70] He negotiated with Nobel about the amount of money to be invested on Sofie's behalf and the form the settlement should take – a pension for life or life insurance. On Nobel's request, he also reported on the conditions offered by diverse insurance companies in Vienna.[71] Although Heinrich Hess adopted a polite, not to say ingratiating, tone in his letters, it is clear that he thought of the settlement in terms of a legal and moral obligation on Nobel's part. The idea that a man must compensate a woman for the loss of virginity if he is unable or unwilling to marry her has a long history. Anchored in the Old Testament, it was taken over into secular law and kept on the books in Germany and Austria up to the twentieth century.[72] Sofie's father thus considered himself an honest broker between his daughter and her ex-lover, whereas Nobel apparently

---

67 Letters 90, 110a note, 164.

68 He bought her a villa in Döbling near Vienna in 1888 (see Letters 129, 139).

69 See, for example, Letter 74. Glenn T. Seaborg lists Gogol, Turgenev, Dostoevsky, and Tolstoy among Russian writers of interest to Nobel (*The Scientist Speaks Out* [Singapore, 1996], p. 133). Ake Erlandsson (*Alfred Nobels bibliotek: En bibliografi* [Stockholm 2002]), however, shows that Nobel's private library contained works by Gogol, Pushkin, Tolstoy, and Turgenev but not Dostoevsky. For a complete list of the books in Nobel's library see http://www.nobelprize.org/alfred_nobel/library/fiction-ru.html.

70 Letter of 28 July 1887, Appendix, p. 278.

71 In letters of 13 April and 14 May 1890.

72 Termed *Kranzgeld*, the compensation was reserved in modern law for demonstrable financial damage rather than loss of honour. The law was abolished only in 1998. See http://austria-forum.org/af/AustriaWiki/Verl%C3%B6bnis.

suspected him of engaging in sharp practices. Heinrich Hess protested such insinuations: "I am not trying to profit from this affair, whatever you may think about me – my conscience is clear, I swear."[73]

At the beginning of 1891 Nobel heard rumours that Sofie was pregnant by another man. She confessed to her infidelity and in July gave birth to a daughter, Margarethe. Nobel declared that their "relationship was at an end" (Letter 161a), but this turned out not to be the case. They continued meeting[74] and corresponding, although their letters now focused almost exclusively on financial affairs – requests for money on Hess's part and protestations on Nobel's part that his generosity and unselfishness were not being sufficiently recognized.[75] His complaints seem specious in view of Hess's abject groveling. In one letter she describes herself as "stupid, naïve, and mindless"; in another as "lacking all reason, idiotic, and a great ass." Conversely she praises Nobel to the skies: "Ah, there is no man like you, and as long as the world lasts, there will be no other man like you. Apart from your noble mind and your benevolence, there is no one who is as gentle and sensitive toward a woman as you are."[76] Rather absurdly, she asks Nobel to approve her marriage to Nikolaus Kapy, the father of her child, and later on even claims that Nobel "obliged" her to marry him.[77] In Sofie Hess's view, she had given Nobel her unconditional love: "You know best how devoted I was to you, how I respected you, obeyed you. Your wish was my command, always … I love you genuinely, with my whole soul."[78]

The relationship between Nobel and Hess will strike readers as dysfunctional and worthy of a Freudian analysis. Modern sensitivities are offended as much by Nobel's chauvinism and bigotry as by Hess's continual self- abasement. One cannot help feeling troubled by the manipulative nature of their letters. Nobel clearly wanted to make his lover feel guilty and unworthy of his "noble actions." She in turn fawned on

---

73 Letter dated 1 December [1889] (= Letter 15 in Arkiv ÖI-5).
74 See Letter 215.
75 See Letters 195, 198, 205, 211.
76 SH 17, 18, 19.
77 SH 39 of 1895. See SH 13: "Kapy wants to marry me … Do I have your approval? You are everything to me in this world. Therefore I beg you to give me your opinion and tell me how you want me to proceed." Compare SH 27 ("I found out that you have no objection to my marrying Captain Kapy") and SH 30 ("I do not know what your plans are concerning my marriage to Kapy").
78 SH16.

him in order to extract money and in later years harassed him with endless begging letters. The domineering language in Nobel's late correspondence with Sofie Hess is in striking contrast to the respectful tone
he adopts in letters to Bertha von Suttner. The discrepancy in Nobel's
tone and attitude towards the two women does not imply that he was
posing, but is situational and shows, moreover, that they appealed to
different facets of his personality. Suttner called forth the idealist in
Nobel; Hess spoke to the authoritarian in him. The two sides of Nobel
are perhaps best characterized by Suttner in "Memories of Alfred
Nobel," published a few weeks after his death in a Viennese newspaper: "His great love for the abstract ideal human being was mixed with
a great deal of loathing, bitterness, and suspicion for real people."[79]
Nobel himself acknowledged that he was "the greatest misanthrope …
but also a boundless idealist."[80]

Ragnar Sohlman, who was close to Nobel in the last few years of his
life and later became his executor, lamented the fact "that in the public
estimation [Nobel] should have figured so much as a rich and remark-
able man, and so little as a human being."[81] It is hoped that this edition
of Nobel's private correspondence will serve as the missing link and
give a human dimension to the public image of Nobel as inventor, businessman, and philanthropist.

About This Edition

I have arranged my translation in two sections: Part 1 contains Alfred
Nobel's letters and Part 2 Sofie Hess's letters. It would of course be
preferable to interweave the letters of the two correspondents to produce a continuous narrative, but this is not feasible because Hess's first
dated letter is from 1891, that is, the series of her letters begins just as
Nobel's letters taper off. By 1895 the numerical ratio is completely lop-
sided: one letter from Nobel to sixteen letters from Hess. That year
Nobel informed his lawyer that he no longer wished to communicate
with Hess[82] because he found her begging letters importunate and dis-
turbing to his peace of mind.

---

79 *Neue Freie Presse*, 12 January 1897, p. 1.
80 Quoted in Fritz Vögtle, *Alfred Nobel* (Hamburg, 1983), p. 95.
81 Sohlman, *Legacy*, p. 42.
82 See SH39 and note.

The dating of some letters in the correspondence is problematic. In the first part I largely follow the sequence and numbering in Arkiv ÖI-5. The most obvious inconsistencies have been corrected by Vilgot Sjöman in his Swedish translation of Nobel's letters.[83] I have adopted his corrections and added some of my own. The dating of Sofie Hess's letters presents even greater difficulties. Sjöman included about half of them in his translation, slotting them according to their relevance to Nobel's letters, but without assigning them numbers. I have significantly changed the sequence of Hess's letters as they appear in Arkiv ÖI-5 and renumbered them to present a more cogent chronological order.[84]

---

83 Sjöman, *Mitt hjärtebarn*.
84 For an overview and comparison of the old and new arrangements see pp. 227–8.

Alfred Nobel, 1883 (© Nobel Foundation)

A young Sofie Hess, date unknown (Vilgot Sjöman, *Mitt hjärtebarn*)

An older Sofie Hess, date unknown (Vilgot Sjöman,
Mitt hjärtebarn)

PART 1

Nobel's Letters

1

[Vienna, 1877]

My dear sweet little child,

It is past midnight. Finally the directors have ended their third session of the day. It is the old story you know about that engages us to this extent. Tomorrow I'm off to Pressburg,[1] where my presence is required. I found the whole business here terribly neglected, and should have stayed longer if it had not been so hateful to me. But I feel *very uncomfortable in the company of these gentlemen*[2] and long to be away from here and back there. As soon as possible at all I'll telegraph or write to you when and where we can meet. In the meantime a thousand, heartfelt greetings and all the best,

Your devoted friend, Alfred

2

London, Westminster Palace Hotel, [16 May 1878]

Dear child,

It is 1:30 a.m., and it is only now that the directors of the company who have tormented me all day long with business talk have left me alone. Never was I more in need of peace, and never have I felt a greater longing for it than today. Write a few words about your health to put me at ease, and you will give me great and true pleasure.

Your Alfred, devoted to you in his heart.

I have such a headache that I can barely see well enough to write. I hope to be in better shape for work tomorrow. Good night!

---

1 In 1873 Nobel established a factory near Vienna in Pressburg, now Bratislava, Slovakia.
2 Italicized phrases in the text are underlined in the original letters.

3

London, Westminster Palace Hotel, 17 May [1878]

Dear child,

Your little letter has been delivered to me and gave me great joy. Write and tell me how you spend your day, where you go, when you drive out, what you buy, etc.

Here I have difficulties finding even a moment to write a few words to you. Meetings with lawyers, negotiations with the minister of internal affairs, scientific debates, and directions for changes to the factories fill my day so completely that we start at eight in the morning and go to bed at around 2 a.m.

Tomorrow evening the negotiations will be concluded, I hope, at least as far as it can be done in the short period of time. Unfortunately bad news reached me this morning of an explosion in the Scottish factory,[3] so that I am forced to depart and go there.

The time drags for me, especially because my headache gets worse and worse on account of the bad diet and the bad air. I will therefore be quite jubilant when I am able to return. When I say *therefore,* you will know of course that I have in mind quite different reasons.

Take good care of yourself, dear child, and don't worry.

Farewell from my heart,
Alfred

---

3 The British Dynamite Company Ltd was established in Glasgow in 1871. The Scottish factory was located in Ardeer and equipped in collaboration with Alarik Liedbeck (1834–1912), Nobel's old friend and chief engineer. For Nobel's impatience with Liedbeck's deafness, see Letter 8.

4

Ardeer, 19 May [1878]

Dear Sofiechen,[4]

So now I'm stuck in this small hole where the wind howls from all directions and keep thinking of the pleasant hours I enjoyed lately in Paris. How are you, my dear good child, in the absence of your grouchy-bear?[5] Is your imagination spinning golden threads for the future or does your young soul wander through the treasure house of your memories? Or do sweet pictures of the present beguile your hours? In vain I try the art of fortune-telling to get to the bottom of this riddle, but I whisper to you from afar a devout wish for your well-being. Good night!

Alfred

5

Paris, 24 August [1878]

Dear Sofie,

There is so much unfinished business here that I can barely collect my thoughts. Work, work, work is therefore my watchword. And to do my work in peace and without worry, you must send me good news of your health. How are you doing in Schwalbach?[6] Does the water and the baths become you? Are you going on outings, are you anxious, does company drive out your sad thoughts? Those are all questions that you must answer as soon as possible. I often think of you and send you loving greetings.

---

4 In the first six years of their acquaintance Nobel addresses his lover almost exclusively as "dear child" or uses diminutive forms of her name: *Soffiecherl, Sofiechen,* or *Sofferl.*

5 *Brummbär* in German. Nobel frequently uses this self-deprecatory term in his letters to Sofie from 1878–1888 (the last time in Letter 130).

6 A spa near Frankfurt, Germany. Over the next ten years Sofie spends most of her time in spas: Schlangenbad, Wiesbaden, Kreuznach (Germany), Ischl (Austria), Carlsbad, Marienbad (Bohemia, now the Czech Republic), Trouville, Luchon (France), Ragatz (Switzerland). Nobel joins her there, but prefers Aix-les-Bains for the treatment of his own maladies.

Your incorrigible Melancholic[7]

Tell Mrs Schmidt to forward any letters that might come for me to: 53 Av. Malakoff, Paris.

6

[Paris], 25 August [1878]

It doesn't take much perspicacity to figure out who composed the French telegram you sent me. He writes quite well, although he makes some mistakes.[8]

I am very sorry that your health is not better. Perhaps it will be helpful if Dr W.[9] takes the trouble to come to Schwalbach, and that is indeed what was agreed on.

I sit at home most of the day and work, yet the time goes by slowly and I feel very lonely. I have quite unlearned to go into society, and am becoming more and more reclusive.

Look after yourself carefully, dismiss all worry, eat well, and amuse yourself well – that is the most useful occupation for you right now.

Has your new chaperone arrived? Are you pleased with her? Who recommended her to you? Write in a little more detail about these points.

You may not believe it, but I spent the whole evening yesterday at my desk and will do the same today. If you don't believe my words, it may be better for your conscience. I have never presented myself as suitor for your heart and therefore have no obligations. You, by contrast, give me so many assurances that I could almost believe they are true if it wasn't for that gentleman there who writes French telegrams for you.

But I am not reproaching you. Take care and you look after yourself well and recover your health and your fresh mind soon. That is what I wish you with all my heart and greet you lovingly.

Your grouchy-bear

---

7 *Grübler* in German. His depressive mood was constitutional, as he acknowledges in Letters 24 and 39.

8 Sofie defended herself against the innuendo that a gentleman friend had written the letter for her. See Letter 10.

9 Dr Walter, also mentioned in Letter 12.

7

[Paris], 26 August [1878]

My dear, sweet child,

Let them doubt it, but it's true: I am worried. That's how it is. I feel very sorry that you are left to strangers. Indeed I feel unease more than sorrow. Please write about everything to me in detail, everything, how you are and how you spend your time, and whether the French chaperone has arrived, and what impression she makes on you. Is it the woman I saw – the redhead? Do you think she will be a good teacher?

Please explain to me, if you can, who sent the French telegram to me or rather who wrote it for you. This is a sore point between us, which has greatly shaken my trust in you. You must always be truthful to me, my dear child, even if you find it hard.

But I don't want to say anything to you that could spoil your good mood, nor can I report anything cheerful, for I feel so lonely and keep brooding in a terrible way. Tell me, my dear Sofie, that you are still thinking of me in friendship, for it looks to me as if all your thoughts were now in Wiesbaden.[10]

Your first letter, which you wrote after midnight because you had no time earlier, gave me the impression that you wrote in an atmosphere of regret for the injustice you have done me.

But I am still fond of you even if you no longer like me and wish you with all my heart a speedy and complete recovery, and blessings from all sides. Only do not forget decency and human dignity in your rush of happiness. Without them no woman can become a true wife or a true mother.

But I don't want to preach to you and conclude with a heart-felt greeting.

Your Melancholic

---

10 For the spas Sofie visited see note 5.

7a[11]

26 August [1878]

Dear Sofie,

I clearly read between the lines of your second letter that you are doing quite well and that my absence gives you joy rather than sorrow. Things are different with me. My life here is very sad, and I feel more lonely and abandoned every day. Yesterday I went to the Théâtre Français[12] to drive out my hypochondria. There they gave an excellent performance of the *Barber of Seville*, but I could not enjoy it, because I feel more abandoned than ever.

Concerning your two chaperones: that is a very sad coincidence or event. Or was it more convenient for you?

Give me a daily report, dear child, about the progress of your health and how you are getting along with your new chaperone. You know of course that I am always thinking of you and, although you are not truthful to me, you have no truer friend on earth than

Your Melancholic

The sister and brother-in-law of my sister-in-law have arrived here, and I shall have to suffer the torture of hosting them. Nothing is more hateful to me than obligatory invitations. You won't believe me when I say how much I wish I were back in Wiesbaden or Schwalbach or Schlangenbad.[13]

---

11 Copy, not in Nobel's handwriting.
12 Or Comédie Française, founded in 1680 by Louis XV. Rossini's opera *The Barber of Seville* is based on a play of the same title by Pierre-Auguste Caron de Beaumarchais.
13 These are the German spas where Sofie was staying.

8

28 August 1878

My dear Sofferl,

Three days have gone by since I had a little letter from you, and so I am rather uneasy. But perhaps you are well amused and have no time to write; and you are quite right, for one must enjoy life when one is young. Later on, sorrow and brooding rise up and cloud the interior sky.

But try to look after your health in spite of attending entertainments. You have only one healthy body, my dear child, and once your health is gone, life itself isn't worth much.

Liedbeck[14] is here since a few days. He sits with me from morning to evening and forces me to shout continually, so that my own ears are numb. Today, for example, he came at 8:30 in the morning; and just now (9 in the evening) I got rid of him. You can imagine how tired I am. And yet I have to set to work, for I have a mountain of letters before me, which have to be answered. That is why I write only briefly today and wish you with all my heart a good water cure and an enjoyable time.

Your loving and devoted Alfred

9

Paris, 29 August [1878]

My dear, good Sofiechen,

I have time only to write a few short lines and to enclose the photo you wanted. I am working hard on the matters about which I talked to you before my departure, but the coming and going of people leaves me so little time that I no longer know how to manage. Now I have to go at once into the city to attend a meeting of the French company, but I will write immediately after my return, and in the meantime your old grouchy-bear greets you more lovingly than ever. By the way, *monsieur* is spelled with an *n*, not *mosieur,* as you spell it.

---

14 For Liedbeck see note 3.

10

29 August [1878]

My dear, sweet child,

My last letter was so short that I have to write you a few more lines. How is it that the old problem has appeared again? Is it perhaps only a consequence of the month coming to an end? Just be extra careful now. The letter from Miss Bach[15] is very pretty. Have you replied to her or would you like me to do so? Do you like your present chaperone better, or would you prefer Miss Bach? Write more about those points.

Oh, so it was Miss who wrote your French telegram? Oh those men, those men, those men!

I send you my photograph today and a thousand loving wishes for your well-being. I hope that you won't be anxious, that you will soon return refreshed and in good mood, that you won't forget your grouchy-bear, that the cuisine of Schwalbach will improve, etc. etc.

Your telegram does not tell me whether you plan to stay in Schwalbach for any length of time or have made up your mind to go to a seaside spa. If taking the cure there does not benefit you, it might be better to spend the whole season at a bracing seaside spa. In all of this I feel so sorry for you, my poor little sick child. I hope you will recover soon and be happy again like a little bird that has been released and given its freedom.

I am increasingly busy, but hope soon to arrange matters such that I have a little more leisure. Since visiting the Exhibition[16] with Brüll[17] I couldn't find a minute to go there again, and yet I need to inspect a few things there. They say the Exhibition will remain open until 1 December so that you will be in good time to see it.

---

15 She was supposed to chaperone Sofie. See Miss Bach's letter enclosed with Letter 13, written from Benzeval-sur-Dives, a popular watering place. She may have been referred to Nobel by the prominent chemist Marcellin Berthelot, a fellow investigator of nitroglycerine. Berthelot stayed in Benzeval in 1877.

16 The *Exposition Universelle* attracted more than half a million visitors and included a Gallery of Machines. It ran from 1 May to 10 November 1878. Nobel was one of the exhibitors. He also comments on the Exhibition in Letter 15.

17 Unidentified – unless "Brüll" (from German *brüllen*, to roar) is a nickname for the deaf Liedbeck. See Nobel's complaints in Letters 8 and 15 about his loud conversations and the necessity to shout back to make himself understood.

If you are a good child and recover soon I will try to arrange for you to accompany me on my trip to Stockholm.[18] But to do that you must first and foremost regain your health, which I wish you with all my heart. So now, my dear Sofiecherl, don't worry unnecessarily and give your old grouchy-bear the great pleasure of recovering soon.

11

30 August [1878]

My dear Sofiechen,

I wish I could, like my thoughts, fly back and forth and be with you. How much more pleasant and restorative the peace of the country is than life here in the city, where one is bothered and crowded every minute by people for whom one has no sympathy, yet has to salute with a nice bow for courtesy's sake. That gets on my nerves excessively and keeps me from working properly. I am now energetically pursuing my project of innovations and hope that they will be successful. But each innovation takes considerable effort, and you must not believe that they can be just clapped together. You must excuse me therefore, my dear sick child, if I write you only very short letters over the next few days. Yesterday evening, for example, Fehrenbach[19] was with me until 11 o'clock, and back again this morning at 7. Every day here passes as if my house was a hotel. They give me no peace, so I have to do my work outside regular hours.

Just now A. Hoffer[20] is coming again, and I can only send you very warm greetings.

Your grouchy-bear, who is always fondly thinking of you.

---

18 He met there annually with his brothers Robert (1829–1896) and Ludvig (1831–1888) and their families to celebrate the birthday of their mother, Carolina Andrietta Nobel (1803–1889).

19 Georges Fehrenbach, a chemist and for many years Nobel's assistant in his laboratory at Sévran near Paris.

20 Amédée Hoffer (1847–1896), an Alsatian engineer who set up the Nobel factory at Isleten, Switzerland, and was one of the founders of the Société Centrale de Dynamite (1870). He was also involved in Nobel's French company in Paulilles, and later became director of Nobel's company in Hamburg.

12

31 August [1878]

My dear, sweet, good child,

Today, once again, I can only write a few lines to you, for I've only just arrived home (I was in the country to conduct detonation tests) and must immediately drive to the theatre with the relatives of my brother. This is how it goes every day. Today I had visits from Walter (whom you know),[21] from an Italian general and an Italian colonel, and many others. I barely have time to breathe. And in addition I feel so uneasy. I wish with all my heart I could see you again and am continually worried about your poor health. Look carefully after your health and make sure not to catch cold. That would be very dangerous at this time.

I conclude with a loving and heart-felt greeting.

Your incorrigible Melancholic

13

1 September [1878]

Dear Sofiechen,

Too bad you weren't with me last night. *Les Fourchambault*[22] was on the program. Such a beautiful piece and played in a masterful way. But first your health: as long as you don't have your health back, you mustn't think of anything else. Since the water in Schwalbach doesn't help you, the only thing left is to continue with the Franzensquelle[23] and to stimulate your appetite with sea air or in some other way as much as possible.

I suppose I should write to Miss Bach or telegraph her. I enclose her letter with a translation. Your reply depends on whether you prefer Miss Bach or Miss […].

Miss Bach's letter is much more elegant in style than Miss […]. Generally, she gives the impression of a refined lady.

---

21 The physician Dr Walter? See Letter 6.
22 A comedy by Emile Augier.
23 A brand of mineral water.

I would like to know more about your decision because I am pressed for time. As you know I am obliged to travel to Stockholm. My work is progressing well and would go even better if I did not have to receive so many people.

Although your health is very poor just now, you must not brood over it, my dear child. It is the consequence of what occurred just now,[24] and will improve by itself afterwards. In my opinion, local stimulation will only be successful after a general strengthening has been achieved with steady drinking of water from the Franzensquelle. That should be your whole focus now, and not to worry unduly. You must always tell yourself that there are thousands of people in the world who are sicker than you and furthermore have no one to support them.

I just received a short letter from my brother Robert. He is in Pressburg and may perhaps come here for a visit. You will like him, for he can be very engaging and amiable once he begins to trust a person. Very likely he will come together with Emmanuel.[25]

Before concluding my letter, I must ask you once again to promise me not to wrap your soul in black robes. In a short while everything will turn out well, and glowing cheeks and a cheerful mood will announce the return of your good health.

Many, many heart-felt greetings from your

Melancholic

I am only sending you a translation of Miss Bach's letter, not the original.

[Enclosure:]

Maison Evangélique
Benzeval-sur-Dives
(Calvados)
France

---

24 A reference to her menses? For similar cryptic comments see Letters 24 and 33.
25 Emmanuel Nobel (1859–1932) is Alfred Nobel's nephew, the son of his brother Ludvig. For the members of Nobel's family see note 17.

23 August 1878

Honoured lady,

I have just received your letter of 16 August and make haste to reply immediately. I am here on the seashore since the beginning of the
month and had planned to return to Paris around September, but I am
completely at your disposal and pleased to take the opportunity of ex-
tending my holiday. It can only be to my advantage, and you were very
kind to think of me. Unfortunately there was a misunderstanding, and
your letter went twice to Paris before reaching me. I shall therefore wait
for another letter from you in which you will kindly inform me whether
I should fetch you from Wiesbaden or from elsewhere. I assume I will
return via Paris, and would have to go on through Metz and Mainz. If
there is a shorter route, I would ask you kindly to indicate it. I would
depart immediately on receipt of your letter. In the meantime please
accept my sincere thanks and respectful greeting.

Your devoted

Bach

14

2 September 1878

Dear Sofiechen,

Yesterday I received no letter from you, which made me very uneasy,
for I ascribed it to your ill health. But I see from the short letter that arrived today that you are better and are amusing yourself, so much so
that you have no time to write to me other than at midnight. You see
how easy it is for you to forget me and how well-entertained you are in
the company of others. Perhaps that is as it ought to be, but take care
of your health in the meantime and don't start too early on a course of
taking a water cure and baths.

Since you are so pleased with Schwalbach now, it is hardly worth the
trouble to consider a journey to a seaside spa. And you will be quite
able to dispense with Miss Bach.

As for me, I am preparing for the journey to Stockholm. In that time
I imagine you will be courted thoroughly. I suppose that is happening
already, but you keep silent about it, like a clever child.

I don't want to grumble, but I would like to know from whom you learned those precious words "Darling," etc. – from an English lady or an English gentleman or a German who speaks English? Yesterday – Sunday – I was home all day working, and today I have already completed some sketches for apparatuses. But that will hardly interest you. I am sure my life here is much more monotonous than yours. The relatives of my brother are a veritable pain, and I will be highly pleased when I have my peace again.

Write often concerning your health, take care not to catch cold, guard against ruining your stomach as I did mine, and think at least once a day of your grouchy-bear who thinks of you lovingly.

15

2 September [1878]

My dear good Sofiechen,

A few well-meant words, without being preachy.

I often think of you in my solitude – for I have never felt so lonely and abandoned as I feel now that I have changed my social position for your sake. You are really foolish to plague and torment yourself without any reason, when you could live so well. But I suppose no philosophy of life can combat feelings, however crazy. So I have to forgive my little toad if her reason succumbs to her feelings. But you must promise me not to allow yourself to be upset unnecessarily or get excited during that critical time, or it will not be possible to restore to you the most precious gift – your health. In the evening I rarely go out, except sometimes after attending the Exhibition.[26] I sit here alone and read or write. Often my thoughts wander eastward, and with a lively interest I ask myself: What is on the mind of my dear, good child, so far away? Has she made the acquaintance of nice people with whom she can amuse herself, whom she invites and with whom she goes for drives? I don't begrudge you healthy, fresh, salutary entertainment, my dear Sofferl. Here you have too little of that, but whose fault is that?

---

26 See Letter 10.

I am finally rid of Liedbeck,[27] and I am pleased because the need to shout continually at this man has dried out my throat. It was no fun, naturally. He is a thoroughly good man, but seems to be unable to understand that his deafness is a bother to other people.

The Exhibition is very beautiful, but there isn't much to learn from it. A veritable sea of light floods the visitor from inside and outside. Think of the whole lower level filled with steam engines and other machines that have the sole purpose of illuminating the interior of the building. It is beautiful, but in my opinion the whole business is not very practical for general lighting.

You scare me with your account of your diseased gums. One could think you got infected with my scurvy. I have often warned you not to use my toothbrush, which should in any case be the rule for all people. Some physicians insist that scurvy is not contagious under any conditions. Others however have doubts about that. Radishes and grape juice is supposed to be an effective antidote. I make ample use of both here, and although I haven't noticed much effect, it appears that my gums are a little less loose. But the sore in my mouth is unchanged. Robert writes from Vöslau[28] that nothing is so effective against scurvy as a regiment of drinking grape juice. As he intends to take that cure, he would like me to join him, but I don't think much of it and told him I would rather drink grape juice at home than waste my time further in a spa.

I have to end my letter now because my eyes are in bad shape. But before I conclude, I take you into my arms, and wish you all the best from my heart and all the good luck your young and innocent soul deserves. With these wishes I send you a heartfelt kiss.

Your Alfred

---

27 For Liedbeck, see note 3.
28 A spa town in Lower Austria.

16

[Middle of September 1878]

Dear child of my heart,

The moment I returned here, I was engulfed in the eternal business affairs, visits, and troubles. Today another relative of my brother was here, depriving me of time from 9 a.m. to 2 p.m. Then I immediately hurried off to inquire about your Franzensquelle, and to my regret confirmed that this brand of water is not available in Paris. If I put in a rush order, it will come from Germany to Trouville[29] in eight or ten days at the earliest. What am I to do now? Will you follow my advice and regularly take the pills the apothecary prescribed for you there? And for your stomach drink cold water from Carlsbad or Marienbad,[30] at least until it is possible to obtain water from the Franzensquelle.

I have not had time to write to Dr Walter – I am under such pressure here – but I will do it in the morning.

Not a syllable from you. How is it with your health and your anxiety? Are you industrious in your studies and are you willing to please your old uncle? Here earth and sky is covered by a grey veil, and there are high waves, but they are the waves of a human flood, or rather the biblical flood. Such a roar after the sacred peace in beautiful Trouville. How a man can change! Years ago I longed for the big city and wanted to see and participate in the bustle and the activities of people. Today, I long to be far away from it and to enjoy sweet earthly peace, a forerunner of eternal peace.

Write in detail how you are, what the doctor says, how you bathe and what medicine you take, how the cook works out, whether you have taken on her daughter – tell me everything, everything that concerns your well-being. And whether anxious guests drive out your anxiety. The best cure, however, is work. Keep up this cure diligently.

Today I was at Valérie's.[31] The portraits are awful, so bad that I don't want to send you any of them. Only the large ones are finished. The smaller ones will likely keep the gentlemen busy for a year. The way they conduct their business is pathetic.

---

29 The seaside resort in Normandy where Sofie was staying.
30 Popular resorts in Bohemia, now the Czech Republic.
31 Portrait photographer in Paris.

Of your old prescriptions I send you whatever I could recover (the one prescribed by Walter, with bismuth and morphine).

A blessing on you, my dear child of sorrows. Do everything you can to regain your health and to study well. With these thoughts on my mind, I give you a tender kiss and remain your loyally devoted Melancholic.

Inform the post office of my forwarding address. There are important letters en route, which I expect impatiently.

17

16 September 1878

My dear, good Sofiechen,

If I sent you only a telegram yesterday, it is because I have to depart early on the day after tomorrow and I can barely manage to finish my work. They have burdened me again with the lawsuit Krebs. I must wade through a whole book and reply, and I have been working on it for two nights now. In addition rheumatism is attacking me from all sides again and giving me a terrible headache. Yesterday I was in such bad shape that I had to go to bed at eight o'clock in the evening. Unfortunately Fehrenbach[32] is also unwell, exactly at a time when I need his help.

I have received all your dear letters, although they came very late. It appears that the mail is slower from Trouville than from Schwalbach. Continue to address them to Paris, where they will immediately be forwarded to me.

Your rural physician may turn out to be quite skilled and may in the end be more successful in effecting a cure than the great doctors.

So far I've had no time to drive to Diculafait and exchange your coat. Nor did I manage to go to Savarre.[33] It is at present very difficult for me to spare an hour. I received a telegraph from my brothers. They have already departed for Stockholm, and I cannot stay here much longer either.

---

32 See note 18.
33 Jules Clovis Diculafait, garment manufacturer; Savarre, Pothet, & Poret, dressmakers on Rue de la Bourse.

In spite of much work, I think of you with loving care and find that you do not tell me nearly enough about your health. If the cook is really so bad, I could let her go if you pay her 60 francs and go back to taking your meals at the hotel. The main thing is to take good care of yourself and make sure of spring blossoming again in your soul and body. It is nice to hear that you study diligently and want to please me. Buy yourself a novel, for example, Balzac's *Deux jeunes mariès*.[34] It is written in the form of a series of letters and in language that is easy to comprehend.

Don't do too much walking until you are completely restored, and don't worry about your health, which is weakened rather than seriously disordered. Keep down your anxiety by thinking that we will meet again soon. Don't eat anything that might harm you, drink a lot of milk boiled with mint leaves and mixed with water from Marienbad, which is at any rate rich in iron. These are the most important rules imposed on you by your old grouchy-bear, who always keeps you in his loving thoughts.

I am in a terrible moping mood, and therefore sign off only with a brief greeting,

Your Melancholic

I enclose the small key. You will find it in the folder for cards. I have safely put away your jewellery.

18

17 September [1878]

My dear Sofiechen,

Why do you damage your health with useless worry? Don't you see how much good will I have towards you and that I show it at every opportunity? One thing is strange: that you cannot understand that I am no admirer of women. I long only for peace, while you imagine that I court women left and right. Can you now get rid of those foolish ideas and return to a healthy common sense?

---

34 An epistolary novel by Honoré de Balzac, published in 1842.

Yesterday evening I invited the Combemales[35] to accompany me to the theatre. That is why I received your telegram too late and could answer it only this morning. In any case you can't expect a reply in the evening since the local offices do not send telegrams after 8 p.m. One would have to take the extra step of sending them to the city.

I won't thank you for that childish business of the biscuits, although it was well meant. The domestics must have had a good laugh at the thought that someone sends biscuits from Trouville to Paris.

Today I am expecting H. at breakfast and P. Barbe[36] at noon. That will pretty well waste the whole day, and I will have only the night left to do some work.

Tomorrow morning, if I can finish my correspondence and the packing, I will depart, but I'll write you a few lines before I go.

As for your crazy suspicions that I will be travelling in company, it is not worth talking about, and yet you are right: I have a travel companion – my powerful and unpleasant rheumatism.

A thousand fond greetings from your

Old grouchy-bear

P.S. If you mix water from Marienbader Kreuzbrunnen in equal parts with water from Forge and Brazza, the resulting beverage will have the same effect as the Franzensquelle, for it has exactly the same chemical composition. You can trust me on that. Indeed, the water from Marienbad has a little more iron than the Franzensquelle.

19

Cologne, 19 September [1878]

My dear, sweet little child,

Since I had to stop here for a day to inspect the d[ynamite] factory near Cologne,[37] I have time to write a few lines to you before the train

---

35 Frédéric Combemale was one of the founders of the Société Centrale de Dynamite in France.

36 Francois-Paul Barbe (1836–1890), Nobel's chief business partner in France. He became involved in criminal speculations and entangled Nobel in financial and legal difficulties. See Letter 148.

37 In Schleebuch, established in 1872.

departs. It pained me to see you so upset and impatient and worrying unnecessarily. That harms your health, the improvement of which should be your first concern. How did you get the foolish idea into your head that I would travel in company? Is our acquaintance so new that you do not understand that I value solitude more than anything else?

Before my departure I sent you water from the mineral source Boulon. It has the same effect as the Franzensquelle, but you mustn't drink more than two thirds of the water in the bottle because the rest is clouded with iron deposit. Keep drinking it regularly to make the beautiful, healthy colour return to your cheeks.

My stomach is once again in such bad shape that travelling becomes a very painful business. They packed up chicken and other things for me (this time I decided myself on what I wanted to take along), but I have no appetite, and the food does not become me. I like only your little biscuits, probably because you meant well sending them to me.

I am sorry to hear that you and Miss Bach[38]are not attending any entertainments because that would distract you and help your recovery.

Tomorrow (for I have to travel through the night) I will arrive in Hamburg, and in the evening I will travel on to Stockholm. The journey is very uncomfortable and tiring, since my rheumatism does not allow me to relax.

Write in detail how you are, for you know how much that occupies my thoughts, and give me soon the much-longed-for news of an improvement in your health. Also tell me how you go about your studies and what you have learned, and see to it that your progress gives me joy. Farewell, my dear good child. My compassionate soul wishes you all the best, from the heart, and my thoughts fly over there to you.

Your Alfred

---

38 See Letter 10.

20

Hamburg, 20 September [1878]

Dear sweet child of my heart,

Before I depart I want to write a few more words and say how sad I am to have no news from you. If you have written to Paris in the meantime, the letter will reach me shortly after my arrival in Stockholm. I often wonder how you are and am not without worries about you. Why I take such an interest in your life is unclear to me – probably because I have such a hard and pitiless heart.[39] Take good care of yourself and write often to me. A thousand heartfelt greetings from

Your old grouchy-bear

21

Stockholm, Sunday [September 1878]

My dear sweet Sofiechen,

A boatload of relatives has gathered here, vying for my time, so that it is quite impossible to find a free quarter of an hour. You must excuse me therefore when I send you only a few short, heartfelt words. Being so far away from you, I am even more worried for my poor, dear, sick child in a distant place. Write in great detail how you are, whether you have received the Boulon water, and how it becomes you. It contains a lot of magnesia and iron and should have as ample an effect as the water from Franzensbad.

My health is not very good. I have a continual headache and nose-bleed so that the least effort is strenuous for me. Here we go: another drop on the paper.[40] Perhaps that will tell you how concerned I am about my little child of sorrow. I feel what Goethe calls *unaussprechlich*.[41]

---

39 Meant ironically, in response to Sofie's allegations that he had a "heart of stone." See Letter 24.
40 A hoax. Chemical analysis showed that the drop is not blood.
41 "Ineffable." Perhaps a reference to a line in Goethe's *Faust*: *Was unaussprechlich ist, sich hinzugeben ganz und eine Wonne zu fühlen* (the ineffable – to devote oneself completely and feel bliss).

Farewell, my dear Sofiechen. A thousand sincere and heartfelt greetings from

Your Melancholic

P.S. I'll depart on 2 October, or if possible, on 1 Oct.

22

Stockholm, 25 September [1878]

Dear child of my heart,

It is a great joy for my old mother here to see us all gathered together. And although I have never experienced great joy (only deep sorrow), I can easily sympathize in this respect with the feelings of other people. When life wanes, and one is only a short step away from the grave, one does not form new bonds and moves even closer to the old ones. I wish I could stay with my mother, and am sure that she would like it too, but we are too many to move in, and I can't make a special case for myself. Thus we all stay at the Grand Hotel, where the furniture is gilded and the food rotten.

I am greatly pleased to see from your letter that you are a little better, but unfortunately also that your company depresses rather than improves your mood.[42] That is a shame and lamentable, for there are people who imagine rigour to be pleasing to God and expect to be rewarded later in heaven with a special, sweet cake. May God keep us from rattling such a beautiful dream. What's really bad about it is the boredom.

From my window I see the flowing river and a small boat, and I wish it could carry me to Trouville. But since that is impossible, my thoughts have to complete the journey on their own and they will remain with you loyally always, now and in the future.

I wish you all the best with my heart and press a fond kiss on your sweet mouth and your little blond head. My soul salutes you,

The old Melancholic

Here the weather is cold and rough, and my health is poor.

---

42 A reference to Sofie's chaperone, Miss Bach, who is with Sofie in Trouville.

23

Stockholm, 26 September
[1878, dated 16 September in a duplicate]

My dear sweet little child,

The hour of mailing letters has come, and I have had no time to write a few words to you. You can't imagine how I am swarmed by people who want something from me and have business proposals. But my brother is worse off. He hardly has a free hour to spend with our mother.

We are quite a crowd here: two sisters-in-law with five of their children. I don't know yet how many of them will travel with me to Paris, but some of them will certainly come, my brothers at least and two nephews.[43]

I am not so fond of the women travelling with us because I won't be master over my time.

I think I may conclude from your last letter that your health is better and you are over your anxiety. So you will soon forget me, dear child. But do keep me in fond memory. I would like to say a great deal more about that subject, but I can't do it today because my nephews are running around in my room and don't even leave my desk alone. My thoughts remain with you and sweeten the boring hours with pleasant memories.

A thousand heartfelt greetings, my dear child. I am writing these cobbled-together lines so as not to let the time for mailing letters pass, and I will write more tonight when the others have gone to bed.

Alfred

---

43 For Nobel's immediate family, see note 17.

24

Stockholm, 27 September [1878]

My dear, good Sofiechen,

Yesterday I had no letter from you. Since you are so far away, I am uneasy and worried. I know that the time is near when every delicate health is severely tested. I hope the weather is good at least. Even here the sun is shining, and it is quite warm considering how far north we are.

You complain, my dear child, about my letters to you being so short and written under pressure, but you don't want to acknowledge the reason while I am unwilling to be explicit. The reason for your attitude is the selfishness of all human beings, and especially of women, who consider only their own person. I noticed it from the beginning and can still see it, and am increasingly sorry that your position in life is awry. I therefore often force myself to be cold and reserved toward you, lest your affection takes deep roots. Perhaps you believe that you are fond of me, but it is only gratitude and perhaps respect, and that feeling is not enough to satisfy the need for attachment and love in your young soul. The time will come, and perhaps soon, when your heart will love another, and you would then wonder at me if I had taken you in my heart and tied you down with deep, heartfelt, and inseparable ties of love. I feel and understand all that very well, and therefore reason forces my sentiments back into narrower confines. Don't think for that reason that I have a heart of stone, as you write. Like others, and perhaps even more than others, I feel the weight of being abandoned in a desert of loneliness, and for years I searched for someone whose heart could join mine. But that cannot be the heart of a woman of twenty-one,[44] whose outlook on life and whose spiritual life has little or nothing in common with mine. Your star is rising on the horizon of destiny, mine is waning. Your youth shines in all the colours of hope, in my case the few shining rays are those of the glow of sunset. Two such creatures are not suited to each other as lovers, although they can be and can remain good friends.

---

44 Nobel was misinformed about her age. She was twenty-seven at the time.

What worries me is your future, for if you would really lose your heart to a young man, and he lost his to you, your faulty attitude towards life would stand in the way of your happiness. I know you don't care for the opinion of others, which is a blessing and will many times in your life save you from getting hurt, but no one is independent of other people's opinion. Self-respect without the respect of our fellow men is like a jewel we can't bring out into the light of day.

Thoughts like these torment me often and often make me moody in your presence and thoughtful and sad in your absence. I know you are a good and dear little child, and even though you have caused me so much sorrow and continue to do so, I am very fond of you and think more of your happiness than mine. Indeed, I can't keep from laughing when I speak of my happiness – as if that was compatible with my nature, which seems created for suffering. But life is smiling on you, my little child, as is rarely the case, and if a little illness comes in between, it will be of short duration. Afterwards you will be joyous and cheerful again. But for complete happiness you lack the necessary culture and education that is appropriate to your station in life, and you must work hard towards it. You are still a child and don't think of the future at all, and so it is good that an old, attentive uncle watches over you.

Just now your letter of 23 Sept arrived. I see from it, although you do not say so, that you are well entertained and that your thoughts are dwelling on other things while you write to me. Reading through half a page, I can't make sense of your words. Perhaps that is the reason why you can't get along with Miss B[ach]. She will not be pleased to see you make so many new acquaintances with men. Don't take my words as a reproach – I am merely trying to find a reasonable explanation for what you are telling me.

Enough grumbling. Take care of your health and be careful, especially now, and don't let flirting tempt you into going out a great deal and staying up too late in the evening. And do not force yourself to write long letters to me, for your last letter clearly indicates that you are forcing yourself.

Here comes my sister-in-law with her children, and I must conclude my letter with a loving kiss from

Your old Melancholic

How are your finances?

27 September

I am staying here until 1 Oct in the evening. If you send a telegram, address it to: Grand Hotel, Stockholm. But add my first name because my brothers are staying here as well.

25

Stockholm, 28 September [1878]

My dear, good child of my heart,

I have such a profound headache today that I can write only a few lines to you. But I must complain, for I receive a letter from you only every second day. I prefer for you to write briefly, but daily. Then at least I know how you are.

Here the food they serve is very bad, and since I cannot take my meals alone, I have continuous stomach problems. Being careful doesn't help, for I do not smoke or drink wine, and yet I suffer such headaches that my whole life seems bitter to me. I hope you are better off. I believe I can tell from your letters that you have already completely forgotten me. That is not nice of you, for you live on in my thoughts and in my heart as always, and I wish you all the best, and send you these wishes from far away together with my heartfelt greetings.

Alfred

26

29 September [1878]

Little child of my heart,

Just now I received from you such a sweet, dear, cute, little letter, dated 24 Sept, that I am very sorry I can't answer it right here and now with a loving kiss. But unfortunately that is not possible, and I cannot tell you even with my pen what I want to say, for I am on the point of departing for the factory and will spend all day there. There is a tumult and ado here such that all day I long for our sweet peaceful time together in […] or Wiesbaden.

If at all possible, I'll write today, but I don't know whether I'll come back in time to do so. Yesterday we were invited to my cousin's. The people were very kind, and yet I felt bored all the time with my surroundings and anxious about what is far away. My head and my stomach are in worse shape than ever. I can't take care of myself here at all.

A thousand, indeed thousands of heartfelt and fond greetings and kisses from

Your old grouchy-bear

27

30 September [1878]

Dear, sweet little child,

Today is the birthday of my mother, and you can easily imagine why I have no time to write many words to you.

Tomorrow I depart – there is an early train – and on 4 October I have a business meeting in Hamburg. On 6 or 7 October I'll arrive in Paris, more likely on the 7th because I suppose I will be forced to visit the factory near Hamburg[45] before I depart.

Perhaps it is not right of me to say that I long to get away from here. I suffer so much from all the dinners and lunches, that I long for France for two reason, first and most of all to see you again, my dear sweet Sofiechen, and also to look after my health, which is in bad shape.

Farewell from my heart and a thousand greetings and kisses from

Your old grouchy-bear

---

45 In Krümmel.

28

The Midland Grand Hotel, London, [10 November 1878][46]

My dear Sofiechen,

My health is bad, and yet I am forced to spend more time in this sad hole. I have meetings with lawyers, one of whom has not arrived yet and keeps me waiting for a long time. I never longed so much to be home as I do now. You cannot imagine the sad life I lead here in England, and on top of everything I am suffering, my stomach, which was a little better recently, has reached an awful low.

And how are you? I thought you would write or telegraph since I gave you my address in yesterday's telegram, but no, not a word. Amen.

A thousand heartfelt greetings and kisses from

Your loyally devoted Melancholic

29

The Grosvenor, 91 Buckingham Palace Rd., London, 2 June 18[79]

My dear, sweet little child,

A little word before I depart – I don't have time for more. After a fairly uncomfortable sea voyage – not much wind, but high seas – I arrived in London at 7 o'clock in the evening. Of course it was pouring as from a sieve or Keller's shower, and of course it is pouring again today. The sky is like lead and the ground a mixture of water and sludge. Looking at that, everyone can understand why the poets sing so fondly of May and June and praise them so highly.

I had a lengthy conversation with the gentleman sitting across from me on the steamer, a very amiable and cultured man. By contrast, [another man], a German, was very self-important and, judging by the whole impression he made on me, associated mainly with easy women because every time the subject changed, he came back to that point.

---

46 The envelope is addressed to Sofie Hess at 1 Rue Newton in Paris, where she lived in 1879/80.

And how was your journey to Paris? Tell me in a few words and don't scribble the address: Alfred Nobel Esq, Dynamite Works, Stevenston, Ayrshire. Scotland.

A thousand heartfelt greetings.

A.

30

The Queen's Hotel, Glasgow, 3 June [1879]

Little child of my heart,

Years ago Bürger wrote: *Hopp, hopp! gings fort in sausendem Galopp.*[47]

You could say that of the express train from London to Glasgow: ten hours covering 700 kilometres, and yet we arrived half a minute before the stated time. In the two wagons that travel direct and require no change, there were only two passengers, a Scot and I. I spent my time reading a historical novel by Victor Hugo dealing with the coup d'état of Napoleon in 1852.[48] It is very well written. And so the hours went by rather fast, although they could not speed up like a locomotive. A good book makes my stay in Scotland bearable.

If you feel a twinge of anxiety or any other female craziness, the only reason is your lack of occupation or lack of society. Buy some entertaining novels therefore and you will surely transfer your anxiety to one of the characters, and everything will come right. Write a few short, loving lines to me, and if you have leisure, try to exercise your mind with some thoughts. Then you will find that everything would have turned out so much better if you had followed my advice from the beginning. But you, on the contrary, almost systematically did everything to alienate me. Don't you know that a sensible woman, if she is so jealous of everyone, need not show it, for nothing in the world is so bothersome as endless crying and endless reproaches and suspicions that recur in the same form again and again like a disgusting dish presented every day.

Lying and cheating is almost preferable, for it offers at least variety. For that reason, my dear child, and in spite of your good and pleasant

---

47 "Hop, hop! On in a rushing gallop," a line from *Lenore*, a poem by Gottfried August Bürger.

48 Perhaps Hugo's *Napoléon Le Petit*.

traits, you make a man's life bitter and disagreeable instead of sweetening it, as a woman should. Think about that. These are words of truth that you cannot understand only because you are fundamentally unable to put yourself in another person's position, and in spite of your soft nature you don't know how to make the least sacrifice.

I have no more time for moralizing – and indeed the effort is wasted on you. Sincere greetings and best wishes for pleasant days until we see each other again, hopefully soon.

Your Alfred

How was your journey back to Paris?

31

Ardeer, 4 June [1879]

My dear, sweet little child,

It is already twelve o'clock at night, and it is only now that I find a free minute to write to you.

So far I got only one short letter from you, dated 2 June, and am glad to see from it that your return journey to Paris went well and without difficulties.

Here in the factory everything is of course monotonous and life unpleasant. But one also has to swallow the bitter pills in life. There are so many issues here awaiting my decision, and it seems that the directors cannot come to any conclusion without my presence.

My work here is going well, and so far I'm quite pleased with my success. But there is such a mass of urgent business that I hardly know where to start.

If I did not have work to do here, it would be the most disconsolate place in the world. Imagine unending sand dunes, uninhabited, lifeless. Only rabbits find a bit of nourishment here and eat what is erroneously called grass and of which there is hardly any trace. It is a strange desert, where the wind blows incessantly and often howls and fills your ears with sand. Sand even floats around in your room like a fine rain. That is the site of the compound. It looks like a large village, with most of the buildings hidden behind sand dunes. A few hundred feet beyond is the ocean, and there is nothing between us and America but water that rises and falls and sometimes rages in giant waves.

Now you have an idea of the place where I am staying. As I said, without work it would be insufferable, but work beautifies everything, and thoughts create a new life that can do without luxury and comfort and never suffers the leaden weight of boredom.

I send you a thousand heartfelt greetings and another thousand from this lonely place, my dear child,

Alfred

32

8 July [1879]

My dear, good, sweet Sofiechen,

There is so much to do here that I can write only a few lines to you today and send you heartfelt greetings. So far I could not deal with the matter of your accommodation,[49] but I hope to have the necessary leisure tomorrow.

I am very tired from the journey and am suffering from a strong headache, which gives me much trouble.

Take good care of yourself, my sweet child, and beware of colds, especially at this time.

When my melancholy thoughts on the journey left me some time, I read *Ein Document*.[50] The story becomes more beautiful and more moving all the time. We will have much pleasure reading it together.

With heartfelt greetings and fond wishes,

Your Alfred

I hope I won't have to stay very long in Paris.
[Address: Madame Nobel, Villa Koppel, Ischl, Salzkammergut, Österreich/Autriche]

---

49 For his interest in renting or buying a house in Peyerbach/Reichenau (about one hundred kilometres south of Vienna) or Hinterbrühl (about thirty kilometres south of Vienna), see Letter 33.

50 A novel by Karl Detlef (pseudonym for Clara Bauer, 1836–1876) published in Stuttgart in 1876.

33

Vienna, 8 August [1879]

My dear little child, rooted in my heart,

Just now your short letter arrived and gave me much joy. I see from it that you understand now why I had to depart and have put aside your bad mood. If you think it over, you will find that I could not act otherwise. It would be unworthy of me to deny my sick mother the joy of seeing me once a year for a few days, as we sons have accustomed her to expect. It would be much more pleasant for me to make the journey together with you. But you could not possibly have managed the strain. In a few days, perhaps even tomorrow or the day after tomorrow, you will not be able to travel at all. Nor would the stay in Stockholm be pleasant for you, for I could barely be with you. We will therefore have to defer our joint visit until the next year.

Yesterday morning I went to see Götz. As you can see from the enclosed advertisement, which is rather old now, he offered his villa in Peyerbach for sale, but the whole family Götz has gone to Zell am See, and no one could give me the least bit of information. Then I drove to Hinterbrühl on account of the attractive advertisement (also enclosed). What a disappointment. It was small, wretched, and barracks-like. And – I agree with you – the Brühl is very pretty, but it is rather rustic. To look at villas in Reichenau isn't worth it, for there isn't a single one for rent that includes bedding.

What shall I advise you to do in the circumstances? Certainly you can't leave Ischl before the 17th or 18th of the month. In the circumstances it would perhaps be better if you stayed a little longer and then went slowly, without exhausting yourself, to Merano, where the season begins on 1 September. Warm weather will become you much better than the wet and dank air of September in Ischl. And in Merano you would get a beautiful apartment right away, perhaps you could even buy a little villa. They say that fall and winter is exceptionally agreeable in Merano.

If this proposal does not appeal to you, I would advise you to journey slowly from Ischl to Switzerland, either Ragaz, Vevey, Montreux or Geneva, where I could pick you up and we could travel together to Aix-les-Bains.[51] I have to do my water cure there if I want to have a somewhat bearable winter. You will not be surprised to hear that I finally want to do something for my own health as well.

If you prefer to go to Vienna instead of staying in Ischl, I have no objections, but I would not advise it because the city is quite deserted, and if there is a great heat wave, it may be unhealthy. But do what you judge best, only make no hasty decisions, as you usually do, and regret them later. Remember your last journey to Vienna, when you disregarded my advice.

It would not look good if I now borrowed more money from Eschenbacher. Therefore I send you a small cheque, which you can cash through Gottwald.[52] Don't write anything on it, for the cheque is made out to the owner. You would do well to give the cheque to Gottwald at once, since it has to be cashed within five days according to French law.

I feel heartily sorry, my dear Sofferl, for your having to live so alone and without support. But nothing can be done about it. One must accept what is unavoidable. How many American women travel almost around the world by themselves! But they know how to get by and are not as helpless as you. And yet I feel sorry for you exactly because you are such a helpless little creature.

Don't do anything unreasonable. Let Olga and Marie[53] take good care of you, don't make nasty remarks when you eat in the dining room,

---

51 Ischl, where Sofie was staying at the time, was a luxurious resort in the Austrian Salzkammergut, frequented by the imperial family. Merano (then Austrian, now Italian) had more clement weather. The expected onset of her menses was presumably the "circumstances" that kept Sofie from travelling. Aix-les-Bains, a French resort, was Nobel's preferred spa. He visited it repeatedly in the hope of curing his rheumatism. For other spas visited by Sofie see note 5.

52 A local merchant. See Letter 68.

53 Marie was Sofie's maid, with whom she had a stormy relationship (see Letters 61, 65, 66, 77, 154). Olga Böttger was an aspiring actress sponsored by Nobel. It was Sofie who introduced her to Nobel in 1881, telling him (falsely) that she was a relative of hers (see Letter 178). In 1895, Sofie accused Nobel of being Olga's lover (SH 39). Nobel denied that his interest in Olga was romantic or erotic (see Letters 172, 178). He told Sofie that he stopped sponsoring Olga when he found that she had lost her moral bearings and "sunk to the bottom" (Letter 183). Nevertheless he left her one hundred thousand francs in his will.

fight against the urge of being overly pretentious. Don't be capricious, don't mangle the German language, and think with a little kind friendship of your protector, who loves you dearly,

Alfred

I should really wait for Trauzl's[54] return to Vienna, but I find that it would take too much time. I will have to depart.

34

Streit's Hotel, Hamburg, 16 August [1879]

Dear, sweet child of my heart,

In all that rush yesterday I couldn't write a single word to you, but I was very uneasy because the expected telegram from you was such a long time in coming. I feel very lonely and sad here among so many vulgar people and I long to be back in Paris where I can work at least or in Ischl, where friendship, devotion, and peace await me. Take good care of yourself, my dear little child, and make an effort to amuse yourself, but an even greater effort to study so that you may make up what you lack.

Today I hope to be able to depart for Stockholm and will perhaps return sooner than you think. Of course I can't leave there too quickly, and will not stay longer than I planned, about eight days. In the meantime you live in the fond memories of

Your Melancholic

My next letter will be dated from Stockholm, and will therefore reach you only in five days.

---

35 is now Letter 78a

---

54 Isidor Trauzl (1840–1929), a chemist who assisted Nobel in establishing a factory near Prague. He eventually directed Nobel's company in Vienna.

36

Westminster Palace Hotel, Victoria Street,
London, Tuesday [1879]

My dear little child,

I have only a few moments to myself, but want to write you a few words anyway. My fingers are stiff from the cold, and I am suffering from such a headache that I can barely see the letters. For people like me with poor blood circulation England is a cursed land. In a few hours I am off to Glasgow or perhaps to the factory, where I hope to spend less time than the gentlemen there have budgeted.

Take good care of yourself, study diligently, eat well and sleep well – but alone![55] I can't give you an address yet since I haven't established yet how long I will remain in Scotland, whether I'll go to the factory or to Glasgow, but I will send you a telegram with details as soon as possible. In the meantime, my dear, sweet Sofiechen, be at ease and don't be anxious; and don't think sad or spiteful thoughts that will undermine your health, and don't forget that health is after all the greatest good on earth.

With a fond greeting,

The old Melancholic

37

Hamburg, 22 June [1880]

My dear, sweet Sofiechen,

I have no time to write you a long letter because here everything is in an awful rush. It is likely that I will have to be present at the proceedings

---

55 Spas were infamous for the amorous opportunities they offered. Palmer, *Austro-Hungarian Life*, p. 128, observes that "a vast amount of mild flirtation" went on in most spas and describes Merano in particular as a "trysting resort" (p. 113). He also reports that Carlsbad and Marienbad were frequented by "Jewish marriage brokers and their clients" (p. 113).

after all, although Bakewell[56] says they will be completed sooner than we hoped. Trauzl[57] is busy with a government commission and unfortunately can represent me only from time to time.

From this you can see, my dear child, that there is work enough, and yet I am sad in my loneliness, and long to be back in your company and enjoy the quiet life at the spa.

Write often and in detail how you are, how the therapy works, and tell me everything good and bad in your life. The former is my heartfelt wish.

Your loving Alfred

38

Hamburg, 23 June [1880]

Dear sweet child of my heart,

I am sitting here alone and abandoned and tormented by unpleasant business rattling my nerves, which are in any case not good, and therefore feel twice as deeply as ever how fond I am of you. The clamour of the world suits my personality less than others, and I would be happy to retire to some corner and live there without great pretensions, but also without worry and suffering.

Once these court proceedings are over, I am determined to retire from my business life. Of course this cannot happen all at once, but I will make a start as soon as it is at all possible.

So far we have no exact date for the court case and I cannot wait here forever. Mr Bakewell will have to make arrangements and manage with Mr Trauzl's help, and summon me only in the most urgent case. I will therefore give written directions to the gentlemen and prepare everything as carefully as possible before travelling to London and Glasgow.

I have an awful lot of work and little time to write at length to you, my dear sweet little child. I can only express in a few words my

---

56 The lawyer William Bakewell (b. 1823), a specialist in patent law, represented Nobel when his patent was challenged in an American court (California) in 1878 (see Letter 40). He had come to Germany to represent him in the Hamburg case as well. Nobel had reservations about him (see Letter 52a), but the case was eventually settled in Nobel's favour.
57 For Trauzl see note 54.

heartfelt wishes for your health. May the treatment make your health blossom. I add a thousand fond greetings from

Your loving Alfred

Always place my letters face-down in such a way that curious eyes can't get a look at them.

39

Hamburg, 25 June [1880]

My dear, sweet little child,

I just received your short letter. You speak of love. How is that feeling possible in the presence of eternal worries and torments? My only love is for undisturbed peace, where no evil men can reach me. Listen to me, and don't waste your young life as you have done until now. Seek the bright light of the sun, not the shadows. Seek happy people, not dark and sorrowful spirits. As for me, there is no joy in this world – I have known that for a long time. I wish you well, very well, and more than well, my sweet child. Use my good will to find a safe haven in life. Then you will be happy yourself, and no disturber of peace will govern your fate.

But you don't listen to my words of admonition, although they are fond words of friendship that call you from afar. Do not chain your young life to my sorrowful existence. Write how you are, in detail, my dear Sofiecherl. A thousand greetings and kisses from

Your loving old uncle

40

5 June 1880

Little child of my heart,

Just don't write anything about the lawsuit to Berzel.[58] That woman is a worse gossip than you think.

---

58 Mrs Berzel is mentioned repeatedly in connection with Marie's dismissal (Letters 46, 47, 61, 66). She seems to have been Sofie's housekeeper in Paris.

Your old uncle is going through a difficult time. These people have come up with a swarm of false witnesses who testify with the most impertinent lies. I shudder when I think how easily a decent man's reputation can be scorched, and how easy it is to deprive him of his money. Luckily there are some survivors who can testify on my behalf.

These are the people who have testified against me so far: D,[59] his wife, his sister-in-law, her brother, an actor in Berlin, two workmen, Heuer, and the widow of Schnell. Quite a battery, isn't it? Very likely more will appear.

In London I could stay only a single night, but left the business in the hands of an old friend who will do his best to bring order to the affair.

Just now your dear telegram arrived, and here is my reply. Make use of the spa as long as possible. I will meet you halfway in Magdeburg or Linz, if feasible. If not, you will have to travel on your own. It is still too cold to go to St. Moritz, and we would have to reserve rooms there in advance.

Today, my dear sweet little child, I can't give you more of my time. I work from morning to evening to develop a basis for the defence. In my thoughts I tell you everything that is affectionate and I joyfully anticipate seeing you again soon. Are you eager to embrace your uncle soon? Just one more little kissy, and amen.

Alfred

Immediately burn this letter because of what I've written above.

---

59 Carl Ditmar, one-time director of the Nobel factory at Krümmel near Hamburg, who left for America in 1870. There he claimed to be the inventor of dynamite and obtained a patent for his product, which he called "Dualin." I have been unable to identify the other witnesses.

41

Hamburg, 6 July [1880]

Dear child of my heart,

Just now I received your longed-for letter dated 5 July, the only letter I have received in many days. The rest are still in the mail, flying in all directions.

I'm happy that you are a little better. I would like to have you here with me, but I don't want to interrupt your treatment if it helps you. I don't have time to travel there. Things are going very badly here. Nothing but my old patent of 1864[60] can save me now. There is a mass of dishonest witnesses. You can imagine how this makes me nervous, given my character. My stomach is at a low, and I can barely eat anything anymore.

How are you supplied with money? – I suppose you still have some. If not, write me in time, that I may send you some.

They have started on your little apartment.[61] But as you know, only a small part has been ordered initially. I have to go back and look after the rest, but when will I find the time for that?

My dear, cute child, I have never felt so bereft of courage. I hope you are in a different shape, and there is only broad sunshine in your young life.

You say you have six days, sweetie, and it would be pleasant to see you again, but I fear the long journey back and forth will fatigue you. I'm not sure what I should advise you to do.

And now I have no more time, my darling. I must go to Krümmel[62] and bid you a fond farewell.

Alfred

Never leave my letters lying around.

---

60 In 1864 Nobel obtained a patent in Sweden for improvements of the process of preparation and use of gunpowder and a patent in England for improvements in the production and use of nitroglycerine.

61 At 10 Avenue d'Eylau in Paris, where she lived from 1880–1884/5.

62 The location of the factory near Hamburg. It was destroyed in an explosion (1866) and rebuilt. The German company Alfred Nobel & Company (later renamed Deutsch-Österreichische Dynamit – Aktiengesellschaft)was founded in 1865 with partners Wilhelm Winckler and lawyer Dr Christian Eduard Bandmann.

42

Glasgow Head office, Nobel's Explosives Co[mpany] Ltd,
Ardeer Factory, 28 November 1880

Sunday

Little child of my heart,

Unfortunately I could not inform you of my safe arrival at the factory by telegram. I did arrive yesterday evening, but at an hour when the telegraph offices in this little village were already closed, and today, Sunday, no telegrams are delivered in this country, not even those of the government.

My hands are so stiff from the cold wind that you will have difficulties reading my handwriting. Such a storm is blowing outside that one can hardly keep on one's feet.

What are you doing right now, I wonder? Are you driving around in Bois [de Boulogne] with your gentle friend or are you sewing at home, quietly and alone? In any case you are better occupied than I who has to feed on indigestible English food, which has made me sick. Life is therefore totally unbearable for me, and all the time I suffer from a headache, and of such intensity that I can't keep my thoughts together. You cannot imagine what torment the least exertion is under these circumstances.

I sent you from London envelopes with my address. Send me a few fond lines and tell me everything, how you are and how you are doing.

Tomorrow I will drive to the other factory. The day after tomorrow I'll be back here. On Wednesday I have a meeting in Glasgow, which promises to be stormy.

I have to stop writing now to visit the factory a second time today. I find that much money was spent unnecessarily in my absence, but it isn't worth talking about.

Loving kisses for my cute little girl,

Your Alfred

43

St. Enoch Station Hotel, Glasgow, 5 December 1880

Child of my heart,

I had planned to depart from London yesterday evening, but the train that brought me here from the factory was so late that I missed my connection, and since trains are not running on Sunday in this god-fearing and god-pleasing country, I am sitting here, lonely and stuck in a hotel so large that it resembles a city district.

When I am forced to interact with people I cannot help noticing how much the lack of a social life in the last three years has harmed me. I feel so stupid and awkward that I am forced of necessity to avoid meeting the people I come across. That is the result of my wretched softness. I will likely never be able to replace what I have lost in intellectual vigour. I am not angry with you, my dear, good, sweet child, for it is fundamentally my own fault, and you couldn't help it. Our ideas of life and aspirations, of the absolute necessity for intellectual nourishment, of our duties as people of higher standing and culture, are poles apart so that it would be in vain to try communicating about these things with you. I am deeply pained and saddened by the decline in the noble aspirations of my mind. Ashamed of myself I retreat from the circle of cultured people. But what is the use of those desperate words? You do not understand me. You understand only what suits you. You are incapable of seeing that I have for years now sacrificed out of purely noble motives my time, my duties, my intellectual life, my reputation which always rests on our association with others, my whole interaction with the cultured world, and finally my business affairs to a child who is wilful and without understanding, who is not even able to see the generosity of my actions.

When I make this bitter confession to you today in writing, it is because my heart bleeds with the shame of having become intellectually so inferior to other people. Do not be angry with me if I complain of my bitter sentiments. You did not know what you were doing when, taking advantage of my compassion and generosity, you undermined my spirit. That is unfortunately how it goes in life: a man who withdraws from all cultured relations and neglects the exchange of ideas with thoughtful people, will finally become incapacitated and lose his self-respect as well as the respect of other people that he has enjoyed.

I conclude, my dear good, gentle Sofie, with the fondest wish that your young life will be better than mine, and that it will never be burdened with the feeling of inferiority, which embitters my days.

Farewell from my heart. Live a wonderful and happy life and think of your wretched and disconsolate friend occasionally,

Alfred

44

Stockholm, 9 June 1881

Little child of my heart,

Only one little letter has come to my hands so far. The reason is the long distance, and I can only hope that you are well and do not worry unnecessarily. On account of a work of philosophy, which I brought along, the journey passed quite well.

You will never guess whom I met in Cologne. At the same time as I, but arriving from elsewhere, a gentleman also booked into the Hotel Nord. The next morning I saw the gentleman only from the back, but I knew him at once: he was my brother Ludvig. Human beings make plans, and coincidence directs us![63] He said he wanted to address a few words to you and ask you to make a small purchase for him in Paris. Of course you will arrange it in the best possible manner.

I felt alright on the steamer, but since my arrival here I suffer greatly from headache and in spite of cold compresses I can hardly scribble a few words. For that reason I keep my letter short, but I wish you exuberant joy from my heart and hope your worries will drop away. Think of Vienna and be at ease like the Viennese children and as your loving old friend desires.

Alfred

---

63 A variation on the biblical "Man plans, God directs" (Prov. 16:9).

45

Stockholm, 10 June 1881

My dear sweet little child,

I just received you dear lines, dated 7 June, with the enclosure regarding the lighting material. Many thanks – the article was very interesting.

It displeases me greatly that your health is not better. We will try the help of country air and good therapy.

My mother is of course overjoyed that I am here and looks ten years younger. Ludvig is expected tonight as well, so that she looks forward to even greater joy. I also prefer Ludvig to his mother-in-law,[64] but perhaps she can be tamed.

Unfortunately I am not well. It is terribly cold here, a very unsuitable climate for me. Be glad, child of my heart, that you are warm in Paris and are not frostbitten, because that is what one is exposed to here. So far no snow, but who knows: tomorrow it may be a toboggan ride. Your boy is cold in the meantime, very cold, and will be glad to turn his back on the north.

Yesterday I spent almost all day at the factory and afterwards the whole evening with Nordenskiöld,[65] until long into the night. Palander,[66] however, is out of the country, accompanying the King. So now you know what is going on here. From my window I see the steamer Gauthiod, which has just made landfall. My brother is on board. I will meet him with my nephew. For today a fond and heartfelt adieu.

Your Alfred

---

64 Ludvig's first wife was Wilhelmina Ahlsell (d. 1869; her mother was Charlotta Elisa-
    bet Asping); his second wife was Edla Colin, daughter of Karolina Frederika Korvin.
65 Adolf Erik Nordenskjöld (1832–1901), descendant of a prominent Finnish-Swedish
    family, a geologist famous for his Arctic explorations
66 Louis Palander (1848–1920), captain of the *Vega*, the ship used in Nordenskiöld's
    Greenland expedition.

46

[Paris, ca. 17 August 1881]

Little child of my heart,

I can only write a few words to you today. Immediately upon my arrival the eternal business affairs started up – soon I'll have to bar my door. Today it started at 8:30 and continued until 1 p.m., sometimes with four people in my room: Combemale, Bilderling, Sparre, Rinière, Hinault,[67] and whatever the others were called. Tomorrow evening I must dine with the Bilderlings. Yesterday I was in Sévran and must go there again today. I am doing the necessary tests there for the Spanish company. My time is fully occupied, as you can see.

The new laboratory is far from finished, but will be very nice and to the purpose.

As for the rest, I feel very lonely especially since I cannot use the laboratory in Sévran as yet and I suffer from anxiety, as you call it.[68]

Write a few words as soon as possible. Tell me how you are, whether the treatment works, and how you spend your time. Better than I am spending it here, I hope. The weather is awful.

I enclose a few lines from Madame Berzel. The envelope was addressed to me. That's why I opened it.

In a few minutes I will go to Sévran. A thousand greetings and another thousand. I wish you a good cure, from my heart, and a pleasant stay. More tomorrow from

Your Alfred

[Addressed to Madame S. Nobel, Grand Hotel Wien, Autriche 43]

---

67 For Combemale see Letter 18. Baron Peter Alexandrovich von Bilderling (1844–1900), a major general in the Russian army, and his brother Alexander were shareholders in Branobel, the oil company set up by the Nobels in 1879. Gustav Adolf Sparre (1802–1886) was a Swedish politician and chancellor of the University of Uppsala. I cannot identify Rinière and Hinault.
68 She calls it *Bange*.

47

Paris, 19 August 1881

Dear Sofferl,

I am writing these lines in the *hamam*[69] where the Arab Ali just gave me a shower and a massage. After such a treatment one feels better for a while.

Most of the time I spend in Sévran where a great deal is still in disarray and my presence is urgently needed. Yet I have found a free hour, as you see, to write to you, whereas you, my ungrateful child, have left me completely without news so far.

But I expected that, for now that you have found some company, you will never think of me again, I suppose.

But I would very much like to know how you are and whether the treatment does you good. How often do you take the mudbaths, and does the Franzensquelle help you? Don't save money, go on nice outings with Olga[70] when the weather permits it. In any case, spend as much time in the fresh air as possible.

I assume your maid has arrived there.

Yesterday I saw the physician who once treated Mrs [Culonne?]. He is said to be exceptionally skilled, and I believe he is. In his opinion I am not suffering from rheumatism.Rather, the physical disorders I am experiencing are the consequence of far advanced scurvy. If it wasn't so late in the season, he thought I should go to Kreuznach.[71] But it's too late now. He tells me to eat radishes as often as possible and after a while he will prescribe strong salt baths for me. He was surprised I'm feeling as well as I do after inspecting my gums and the sore in my mouth.

You can see that your boy isn't so healthy and strong after all.

The letter from Mrs Berzel, which I forgot to enclose the last time, is enclosed now

With a thousand heartfelt greetings,

Your Alfred

---

69 Turkish bath.
70 For Olga see Letter 33 and note 52.
71 German Rhineland spa, forty miles southwest of Mainz.

48

Paris, 21 August [1881]

Dear good Sofferl,

I just received two letters from you and am glad to see from them that you are taking your therapy seriously. It is possible that there is a little something wrong with your liver, but poor blood seems to be your principal problem. I hope the baths will eliminate that problem.

Go on many outings and make sure you always take the air. Let me know in time if I have to send you more money.

I spend most of my time in Sévran, but I do as much as I can to get rid of that wretched scurvy.

How is it that this year there are so few refined and agreeable visitors in Franzensbad whose company you could enjoy? Have all these refined ladies suddenly recovered their health and need no more spas? Or have they found out, after taking all those treatments, that they are not very effective?

Apart from having dinner with Bilderling,[72] I have paid no visits and received no invitations. People hardly know that I am here. No one gets to see me because I spend most of my time in Sévran.

Give my greetings to little Olga and write soon. You know how much I am concerned about the improvement of your health.

Your loving, devoted Alfred

49

22 August [1881]

Little child of my heart,

It is late, but I write a few words to you.

The place here is deserted and abandoned. I wish the equipment in Sévran was finally installed so that I can seriously start my work. My eyes are in very bad shape, so that it is an effort to write a few lines to you.

---

72 For Bilderling see Letter 46.

Take good care of your health and see that you return like a little blossoming rose. Nothing has any value in life when one is ill, and even poverty is rosy when accompanied by good health. Think about that, my dear child, and do not allow yourself to be distracted by gloomy thoughts, which is silly. That is my heartfelt wish.

Your loving, devoted Alfred

50

26 August [1881]

My dear, good Sofferl,

Almost every day I receive a sweet little letter from you. I can see clearly how good it was to allow Olga to accompany you. I can tell from your style of writing that you are less nervous and uneasy than when you are completely on your own. I can't say that your style of writing is very good. That would be a lie, but your letters are more correct and better than ever before. Thanks for the article about the planned Panama railroad. I had already heard about it. It is a grand affair undertaken with the typical American daring. As far as the channel of Lesseps[73] is concerned, there are problems. Workers are dying like flies. A terrible fever rages among them, they say.

In Sévran everything is making good progress. The arrangement is very much to the purpose. I must spend almost all my time there and can't look after myself as I should.

The method used by my new doctor seems to produce good effects. My gums are a little less sensitive, and the sore in my mouth has not worsened. I can't draw any definite conclusions, however. Perhaps the improvement is due to the weather: I haven't caught cold and therefore feel better. There is no improvement as far as my eyes are concerned, and both reading and writing are becoming very difficult for me. Fortunately I receive only few letters these days.

I tell you so much about myself because I know you want to keep informed about the health of the old grouchy-bear, I mean Dr B.[74] On the

---

73 Ferdinand de Lesseps (1805–1894), the developer of the Suez Canal.
74 This is a private joke. Nobel also signs himself "Dr Grouchy-bear" in Letter 53.

whole, however, the subject is of little interest – whether the old man does or doesn't see well, whether he does or doesn't live, has or doesn't have rheumatism – no one gives a hoot, and no one would miss him if he moved into an apartment under rather than above the earth.

How are you, my dear child? Are you still drinking Carlsbad water, and how does it work for you? I don't agree with the doctor who insists that one should drink only from the source. That may be true for mineral water that is very hot, but not for other sources, in my opinion. Adopt the advice of Dr Klein.[75] He seems to take matters seriously and is not completely deceived, I hope.

Would you like me to send on your fur coat? It would be better if I could avoid it because packing it up is a little difficult and I can't be sure that it will be safely delivered. Nor do I think you will have room for it in your luggage. I believe the cold has abated, but if you really want the fur, send me a short telegram.

I enclose something to fortify your account: a thousand francs. The exchange rate in Vienna is about 214, or 467 gulden. There are enough money changers there.

How do you like Marie?[76]

Does little Olga amuse you a little? In any case, it is good for you to have someone keep you company when you go out or drive around.

Farewell! Tomorrow I'll write when I somehow find the time. Today I spent all day at the French company. The discussion was heated, for we have important decisions to make.

With fond greetings,

Alfred

I enclose a letter of Madame Berzel.[77]
[Addressed to Sofie Hess in Franzensbad]

---

75 Not identified.
76 For Olga and Marie see Letter 33 and note.
77 For Berzel see Letter 40 and note.

51

1 September 1881

My dear, sweet child,

I have just received your heartfelt lines dated 28 August. You complain, my dear Sofie, because I don't write in a warmer tone. I keep telling you at every opportunity and have been doing so for years that feelings can't be forced. You have a kind heart, are dear and good, but you make a nuisance of yourself, and my freedom-loving nature does not allow me to lead a truly happy life with such a person – especially when you are suspicious, jealous, and childish. I admit that conditions appear to have improved a little, but even so solitude is preferable at this point.

I have, moreover, become a stranger to society, and have hardly any friends. Therefore I long for an intimate relationship with someone who understands me. But you do not understand me. You don't comprehend that a free spirit does not wish to be chained or tied down. He can put up with chains, but only under the condition that he cannot feel them.

I am sorry, very sorry for you, and all the more so as I get to know you better. You might have been able to tie me down if you had begun by making my life pleasant instead of unpleasant. But you did everything at that time to alienate even the best person. Now your young, gentle, good soul thirsts to be loved in turn, and you are right to find my feelings too flat. But whose fault is that? I always say: try to win the true, enduring, profound affection of a decent man, and establish genuine and true family connections, which are incompatible with a deceptive position. Your ill health is perhaps more the result of a feeling of emptiness and unfulfilled longing of the heart, which weighs you down more than any actual physical disorder. Take a serious look at your life and take my words to heart. What I say is well-meaning, fond, and sincere advice.

But enough sermonizing. How long do you intend to stay there? The weather right now is very cold, but they say it will change in a few days. That is my fondest wish, for otherwise the treatment can only be half as effective.

I keep going to Sévran, and I visit the Turkish bath as often as I can. I also drink iodized radish and grape juice. I don't feel that it is of great use, but it helps somewhat. In any case it is better than hanging about those stupid spas, where one kills time in the most idiotic way. Poor health is a thousand times preferable to intellectual death.

Give my greetings to little Olga. To you I send a thousand and another thousand fond and heartfelt greetings,

Your Alfred

Do you still have some Tokay wine left? Otherwise you might order a
few more bottles from Polngyay.[78]
[Addressed to Sofie Hess in Franzensbad]

52

4 September [1881]

My dear, cute little child,

On the whole the treatment with radishes and grapes that I am taking
seems to do me some good, but for the last two days I've had a cold
again, which takes away everything I have gained.

Since my return I haven't visited anyone, not even Hugo and
Daubray,[79] who looked me up recently but did not find me at home. I
should have returned his friendly visit, but so far I had no time. Today
is Sunday, and I felt so lonely that I decided to drive to St. Germain and
breakfast there. It is quite nice there, but in this cold it is no pleasure to
go for a drive.

And how is my little toad? No letter for two days now! I am not surprised. I suppose my little, gentle child feels low. Be very careful and
don't catch cold. You never answered me about sending your fur coat.
I walk every day and hope for warmer weather, but it does not come,
and I feel sorry for you in the sad town of Franzensbad burdened with
cold weather as well. Olga is diligently studying French, I suppose, and
will surprise her mother on her return.

You haven't written anything about your living quarters. Are they
comfortable and warm, and is your room large enough to allow you to
walk up and down in bad weather?

Write when you are able to start with the baths again. I will then be
able to judge to what extent it is worth continuing with your therapy.

---

78 The word is hard to decipher. This is Sohlman's reading (*Legacy,* p. 67).
79 Victor Hugo (1802–1885), leading French Romantic novelist, poet, and playwright;
    Michel René Thibaut (stage name Daubray), leading French actor (1837–1892).

I don't know whether you can read what I'm scribbling today. I have a sore throat and am running a fever and have hardly slept for the past two days. That makes me so nervous that I can't write clearly. But I *can* wish you all the best from my heart. I conclude with a heartfelt greeting.

Your loving, devoted Alfred

52a[80]

Hamburg, 27 November [1881]

My dear, sweet little child

I'm still stuck here with Mr Bakewell, doing preparatory work. We still don't know when the trial will start. I won't be able to wait for it, and as soon as Trauzl is somewhat *au fait*, I will go on to London and Glasgow.[81]

Mr Bakewell is anxious, a quality that makes him quite unsuitable for dealing with such cases, and he is so indecisive that he doesn't dare hire an assistant without telegraphing New York first.

Until our opponents have produced a list of witnesses, everything is vague. It is only at that point that we can see what witnesses we ought to interview on our part.

Trauzl read through all the account books and said he could not understand how one could pay out so much money in response to crazy demands such as Ditmar's.[82]

So far I have received three short nice letters from you and am pleased to see from them that you are taking your treatment seriously and paying attention to your health. Bury yourself in that beneficial dirty mud as often as the doctor permits it. And don't come with tears and complaints that undermine your health. Don't forget, my dear Sopherl [*sic*]: You have a stomach ailment that must not be taken lightly.

My own health is a little better than I dared expect in the circumstances. It appears that the treatment there did have a positive effect in the beginning. But I also watch my diet and eat very little. Conversely

---

80 Copy, not in Nobel's hand.

81 For Bakewell (here spelled "Backwell") and Trauzl see Letters 33 and 37.

82 For Ditmar see Letter 40 note.

I take cold showers when I have the time, which doesn't happen often, however.

And now, child of my heart, I give you a heartfelt remote kiss,

Your loving old uncle A.

53

8 September [1881]

Dear, good Sofiechen

I was so much better that I already had genuine hopes of regaining my health, but in the last five days I got worse again. A light cold brought all that about. If at all possible, I intend to take the baths in Kreuznach for a few days, to discover at least whether such a treatment can be of any help to me. I have to try something since I am completely sleepless and have no appetite whatsoever.

Your dear little letter of 5th September gave me great joy. I can see from it that you have not completely forgotten your old protector. Olga writes that you do not look well and are much affected by taking the baths.

It is most regrettable that the therapy does not help you as much as I had hoped. You are not richly endowed with good health, and I wish you would lay in a good store for the winter during the warm season.

I have a great longing, my dear little child, to see you again happy and cheerful and hope at any rate that the mudbaths have strengthened you and lifted you up.

Your telegram arrived so mangled that with the greatest effort I could decipher only half of it. I could make out only that you were able to start another water cure yesterday. In that case you will have to be patient and stay until the 15th. I don't know as yet whether I can pick you up. But it is likely that I'll wait for you in Kreuznach or Wiesbaden, which are not far from Franzensbad. For my thoughts and feelings, certainly, it is only a short distance, for they are with you, and in my thoughts I address to you some sweet words and give you a heartfelt kiss.

Your loving and devoted Alfred

Greetings to Olga from the old Dr. Grouchy-bear.

54

[London, February 1882]

My dear, sweet Sofie,

You cannot imagine what a bad crossing we had. The waves rolled over the ship as if it had been a little boat. I wasn't exactly seasick, but felt something even worse, a general malaise that will have consequences later.

Here affairs weren't nearly as well prepared as I had hoped and the gentlemen had allowed me to expect. The general opinion is that the business, if it is to be transacted, can only be done in Glasgow and through the influence of friends. In that context it is especially useful that I am widely known in Scotland as a businessman. I find the whole affair exceptionally hateful, and if I wasn't already so deeply involved, I would keep my distance from it completely. I don't mean to say that I am unwilling to do everything in my power to be useful to my brother[83] and his business in every way – I'm not at all unwilling. But our views diverge in many ways, so that we really should not work together in the same business. I intend only to introduce him to the gentlemen in Glasgow and leave the rest to him, or at the very best point out the direction in which to proceed before I withdraw.

I am very sorry to leave you so completely alone now, but if you knew how much depends on this matter, you would find it more understandable. It is truly amazing that people who are in charge of this giant company are struck blind in financial matters. As far as I could understand from the secretary, my views are fortunately shared by the company in Petersburg.

---

83 I.e., to provide financing for the oil company Branobel, in which all three brothers held shares, but that was directed by Ludvig Nobel. In a letter of 1883 Alfred candidly wrote about their divergent views on how to conduct business: "You build first and procure money afterwards, while I suggest that in future it is best to find money first, and then to enlarge" (quoted in Bergengren, *Alfred Nobel*, p. 83). After expanding too fast, the company was short of capital in 1883 and in a precarious position by 1884, from which Alfred Nobel saved it by arranging loans with the Credit Lyonnais in Paris and the Discount Bank in Berlin. His brother Ludvig, in turn, was given credit by the Russian State Bank. By the end of the year the financial crisis had been solved. See also Letter 72.

But enough of business. I have not received even a syllable from you and don't know how you are. Write a few heartfelt words. Tell me what you are doing, how you are. Take as little medication as possible, and accept a thousand heartfelt greetings and kisses from

Your Alfred

London, Friday.

We have left the Grand Hotel, where we had only two small rooms. My new address is Buckingham Palace Hotel, London.

I enclose two English bank drafts, each worth about 1,260 francs, in sum 2,520 francs. Any money changer will accept them.

55

Glasgow, 28 February 1882

My dear, anxious little child,

Although it is already midnight, I write a few lines that I can however only post tomorrow morning. Obviously you have received my two telegrams, one from Folkestone, the other from Preston en route to Glasgow. The crossing was very bad. There wasn't much wind, but raging waves after the storm we had day before yesterday. You would have died of fear and seasickness.

On the stretch from London to here I was in a large salon-car without travel companions, except for a Scotsman who came on board shortly before Glasgow. You can see that I had much time to brood.

I tried to read *The Refugee's Son*,[84] but as usual when a novel has passed through your hands, a lot of pages were missing, that is to say, pp. 32–97. You should really buy only bound books.

During my stay here in Scotland, my dear Sofferl, you cannot hope for many letters from me. My time is so much in demand that I cannot find the leisure to write, and in any case I am never alone here. People want to entertain me and don't understand that they are only a nuisance to me.

---

84 *Der Sohn des Flüchtlings* (1881) by Valeska Bethusy-Huc.

If you feel anxious, drive out your mood with work. That is the only reasonable treatment for that illness.

And now, child of my heart, I kiss you a thousand times.

Your loving, devoted Alfred, whom you torment so much

56

Glasgow, 4 March [1882]

Dear little child,

I have time only to thank you for your dear little letter and to tell you to admit joy through one window and dismiss anxiety through the other. Life is too serious a business to pass the time only with childish things. Here I am overwhelmed with work. The company is in a very difficult position on account of my long absence, and I must rescue it from a difficult situation. The factory cannot deliver even half of the orders, a nice result that comes from being tormented and chased around for years now by an irresponsible child. But I don't want to blame you, which in any case cannot make up for the past. The great intellectual effort here makes me very nervous. You can probably guess as much from my writing – and I am quite without sleep. Yet I feel more content because at least I don't waste my time and can be productive.

I will write soon. Just don't ask me for a long letter, because my time is always fully occupied, and I barely have a few moments to myself. But I wish you good health and contentment with all my heart and end my letter with loving greetings from

Your fond and devoted Alfred

57

London, 8 June [1882]

My dear, delicate Sofiecherl,

What a difference it makes to spend a day with you quietly rather than working here from morning to evening, and almost from evening to morning, because I often look after business affairs until 2 a.m. As soon as I arrived here, two gentlemen waited on me at the hotel to take me to a club (no ladies are admitted to clubs here). Hoffer, Wayemann,[85] and I returned from there at 1:30 a.m., and I could not fall asleep, so that I spent a second sleepless night. It is a terrible way to live, and my stomach is in complete disorder. Today we all dine at Doctor Dewar's Club. Tomorrow, Sunday, I promised to go to Mr Webb's[86] country seat and spend the day there. Monday Professor Abel[87] has put in for my time, to dine with me at his club, and so it would go on until the Second Coming if I did not forcefully put an end to it.

Business affairs are progressing all right. If one negotiates for fifteen hours, one can show something for it.

It appears that you did not receive the telegram I sent to Frankfurt since it was returned to me. I enclose it to prove to you that I sent you a message as soon as possible. I also send you a letter from Mrs Böttger.[88] It was addressed to me. That is why I opened it. I could not possibly guess that it was for you.

When I departed, I was very sorry to leave you so alone and abandoned in that large hotel, although I had no doubt that Mrs Böttger and Olga would arrive that night. Yet my heart was heavy to leave my little witch so alone without any protection. Write, child of my heart, and tell me how you are in Carlsbad (I know from your telegram today that you arrived there). Are you satisfied with your living quarters? Don't forget the promised five florins for the porter. How does the treatment become you – I mean, how well does the water agree with you, how do you like your doctor, do you see Robert often,[89] how is little Olga, etc. Give her my regards.

And now I end my letter with a loving long-distance kiss and wish you with all my heart a good and complete recovery of your health and all joy in life, which will then come by itself.

With fondest greetings,

Your Alfred

---

85 For Amédée Hoffer see Letter 11; James Dewar (1842–1923) was a chemist who collaborated with Frederick Abel on the development of cordite (see note 86 and Letter 169). I have not identified Wayemann.

86 Orlando Webb, a business lawyer and British import agent for blasting oil from Nobel's factory in Krümmel. He was on the board of several mining companies. See Walter Skinner, *The Mining Manual for 1888* (London, 1888), 645.

87 Frederick Augustus Abel (1827–1902), chemist and lecturer at the Royal Military Academy, Woolich. He was later involved in a lawsuit with Nobel. For the so-called Cordite Case, see Letter 169.

88 Olga's mother. See Letter 33 and note.

89 Nobel's brother. For his family see Letter 10.

58

Paris, 16 June 1882

My dear sweet little child,

Altogether I have received only three letters and two telegrams from you. Your and my last telegram crossed in the mail.

Day before yesterday and yesterday I was in Sévran, but I had such a mountain of letters from England to which I had to reply that I could not get on with my chemical work. Today I am at my desk since 8 a.m. and have not even taken off my slippers.

Apropos slippers, it would please me greatly if you could find my light boots with the blue lining that I left behind at the Kaiserhaus.[90] They are very comfortable boots and suited both for a soirée or for going to the theatre and can't be replaced easily.

It has been a long time since your last letter, which came on 12 June. I long for news how the water agrees with you. In my opinion a dose of three cups a day is much too strong, but the doctor knows best. As long as it agrees with you so that you might have a pleasant winter!

It is very nice of Ludvig[91] that he asked you to write to him, but don't forget that your style leaves much to be desired and make an effort to write briefly, yet mix in something thoughtful.

Here I consulted the well-known spa doctor Const. James, a handsome man.[92] The question is: will his judgment prove correct? He advises very strong sulfur baths for me in Bagnère de Luchon in the Pyrenees. He believes that I will have a pleasant winter if I follow a vigorous regimen there. Well, let's see.

Day before yesterday the famous national feast[93] took place. It is said to have been very grand and brilliant. I was in Sévran and saw nothing. All that is left to see now is a display of many flags and coloured glass lanterns. I am not in the mood at all for sightseeing. I find that the summer is passing too fast without any results, and my conscience bothers me as much as my neuralgias.

---

90 A hotel in Carlsbad.
91 Nobel's brother. For his family see Letter 10.
92 Dr Constantine James, who wrote an article about "miracles" of healing in Lourdes (*Foreign Literature, Science, and Art* 37 [1883], p. 54).
93 Bastille Day (La Fête Nationale), which is celebrated on 14 June.

And how is my dear muddy[94] yet lovely child? Does Olga really offer you good company? And are you strict in following the instructions of the doctor? And does he court you assiduously? And what about your sick little stomach? Is it still bloated like a filled balloon? And are your intestines getting stronger? And your mood lighter? I want to know all about it, and my dear little toad must write to me at once.

I am busy here to obtain a loan for my brother, but all this extra work takes away from my time that I could usefully employ in Sévran. As you can see from my writing, I am nervous again. Work is piling up, and time is getting shorter.

I send you fond greetings.

Your Alfred

59

Paris 17 June [1882]

My dear little child,

I wrote you a long letter yesterday and today I add a few lines. Why don't you write anymore? I am so interested in your well-being. You don't need to spin a long story, just let me have a few words about your health. If only the doctor would not prescribe too much hot water from the source for you! Such drastic cures aren't suitable for you.

I am at the point of departing for Sévran, but will return in the morning and hope to find some lines from you on my return. Or have you completely forgotten me? It would not surprise me. *Ainsi va le monde* [that's how the world turns], as the French say.

I enclose some English money to fortify your cash box: a bill of 20 pounds and six bills, each of 5 pounds. Altogether 50. You will get some 575 florins for them, depending on the exchange rate.

With fond greetings,

Your Alfred

---

94 An allusion to the therapeutic mudbaths Sofie was taking.

60

Paris, 23 June [1882]

My dear sweet little child

How can you be so silly and want to depart now when you have just begun your treatment? Do you think your health is a dispensable commodity that one can throw out without further ado? On the contrary, it is of the first importance and the most significant thing in life, without which all wealth and all jewellery won't be of any use. It sounds odd that I have to lecture you on that, but you are unwilling to use your own brain.

The doctor, whom you abandoned, is not incompetent at all in my opinion and was warmly recommended by Klein. The fact that you consulted Dr. Schnee on my brother's advice gives me pause. In any case the matter can't be undone, but if he prescribes a radical treatment for you, make sure you consult Dr. Klein before you accept it.

You didn't answer my telegram in which I asked you where to send the tortoise combs. It is probably too late to send them to Ludvig since he stays only until the 26th in Marienbad. He says he will look you up before his departure.

My poor little child, you are always worried about one thing or another. In that fashion you won't have a good life and will make others wretched as well. The worst is that you can't afford to risk your health, and go on undermining it with your ill feelings, until neither treatments nor doctors can help you. Consider how young you are and what a beautiful life you could lead. Make an effort to regain your health and return to Paris fresh and blossoming like a rose.

I spent too much time in Sévran yesterday and could therefore not travel to Luchon this morning as planned. Tomorrow morning I will go there. There is no time to lose. At the end of August it is supposed to be quite cool in the mountains, and that would not be suitable at all for my treatment.

I was at the point of writing you this morning, when Mr Barbe[95] was shown in. You cannot imagine how the man has changed, and there is no other reason for it than the death of his mother followed by that of his father. The old man was not at all pleasant, but he was close to Barbe, and offered him advice and understanding. Now he has developed a

---

95 For Barbe see Letter 18 and note.

"spleen," as he expressed it. I tell you all this to show you that only very few people are able to live alone, and your little boy is a rare example, because he welcomes being alone and finds in solitude the peace he always longs for.

Not much new here. Day before yesterday I had to accept an invitation to lunch at Hugo's,[96] as it was the old gentleman's name day. The visitors were the usual crowd, except for a captain who recently came up with a daring project to create an interior sea in Africa and found in Lesseps a supporter for his proposal.[97] A very pleasant and, it seems, modest gentleman.

As for your mad idea of leaving Carlsbad early, I have to tell you off properly, but at the same time send you fond greetings and many heartfelt words.

Your Alfred

61

[Hamburg, August 1882]

Child of my heart,

Just now the waiter brought me your sweet little letter from Ragatz. Unfortunately it appears that you are once again inclined to go into a snit. You must fight it with all your strength, if you want the advice of a man with experience. Believe me, melancholy has a negative influence on your digestion, more so than bad food, and for people with your constitution, good mood and cheerfulness are more useful than four weeks' treatment in Carlsbad. Therefore read pleasant books, or have them read to you, and fight with all your strength against the return of a bad mood, for that prevents almost singlehandedly the recovery of your health.

Enough sermonizing. I finished reading *Das Haus des Fabrikanten*[98] on my journey and will send it to you in Franzensbad as soon as I know your address. The second part is very interesting and moving and you will find the book engaging. I arrived in Hamburg very fatigued. The

---

Victor Hugo. See Letter 52 and note.
The "captain" is François Elie Roudaire (1836–1885), who proposed flooding parts of the Sahara that were below sea level. For Lesseps see Letter 50.
Published in 1883 by Gregor Samarow (pseudonym for Oskar von Meding, 1829–1903).

gentlemen clearly noticed my exhaustion and therefore agreed to move the meeting to the afternoon. Now I am better, and I realize that travelling through the night does not work for me. The trip to Sweden will also be strenuous since I can't hope for nice weather.

I will growl a little longer because of your outing to Ragatz to accompany me. That was very careless of you and no doubt brought on your menses.[99]

The enclosed letter from Marie was delivered to my address. Since I don't know your intentions, I can't reply directly to her. I suppose you will give your orders to Mrs Berzel, if anything needs to be done in this matter.[100] If I understood you correctly, Marie said at that time that she could not show a recommendation from you. If you are sure that you understood her correctly, I would not give her a recommendation if I was in your place.

Make sure that you will benefit from your treatment in Franzensbad, and that you will depart in a few months fresh and blossoming and will look at the world with a light, cheerful heart.

My journey to Stockholm will be torture since Liedbeck is here and absolutely insists on coming along. Imagine three days of shouting and pity me![101]

A thousand heartfelt greetings and kisses.

Your fond and devoted Alfred

Regards to Olga

62

Hotel J. Sorensen, Copenhagen, 12 August [1882]

My dear Sofiecherl,

A large medical conference is taking place here, attended by thousands of physicians from all over the world. As a result I searched in vain for a place to stay. I looked from midnight until 2 a.m. and had to consider myself lucky to move into a servant's room in a miserable attic.

---

99 *Unwohlsein* in German. This may be a reference to a general malaise or, more likely, to Sofie's menses. Ragatz is a spa southeast of Zurich, Switzerland. For other spas mentioned by Nobel see Letter 5 note.
100 For Marie and Mrs Berzel see Letter 40 and notes.
101 For Liedbeck see note 2.

In half an hour I'm travelling on to Stockholm and because I want to eat something first I have time only to greet you from my heart and embrace you.

Alfred

63

Stockholm, 29 August 1882

My dear, sweet little child,

Finally I have arrived here and find my mother, who will be seventy-nine years old, as well as one can expect in the circumstances. The old woman was so glad to see me, her eyes practically sparkled. What do people see in me that they become so attached? The more I think about it, the less I can understand it.

Your short letter from Ragatz has just arrived. What you tell me about those eight days is typical. This isn't a matter to be taken lightly. Yet I believe that your return to Franzensbad was urgently needed, and I counsel you earnestly to follow your doctor's advice with the utmost attention. He is a competent and conscientious physician.

I can't write more today. On this first day of my stay here, my mother naturally occupies my whole time.

My fondest wishes for a good outcome for your treatment.

With loving greetings,

Your Alfred

Regards to Olga

64

Stockholm, 31 August 1882

Dear sweet little child,

My dear mother took such pleasure in my presence that my heart was touched with regret about leaving the old woman on her own throughout the year. Recently, when my brother Ludvig had to depart, it seemed she was so affected that she fell ill and only gradually recovered. At the age of seventy-nine one no longer has an iron constitution, and it doesn't take much to weigh down the body.

Since my arrival here I am unfortunately overwhelmed once again with correspondence. Thus I have to spend many hours writing, which I would rather devote to the old woman. She is like you – when she is happy and cheerful, she can take anything. When she is not happy, the least thing harms her.

Yesterday we drove to the zoo and dined there. The food agreed badly with me – I was sick for the whole night as a result – whereas my mother was quite well afterwards.

How is your treatment going? It is a miserable business to be ailing all the time. Take care of yourself so that at least you don't spend time in that boring hole for nothing.

I hope the weather there is better than here. Wind and rain, rain and wind – that's the only variation we have. Nor is it warm here, I never took off my winter coat.

So far I wasn't able to visit Nordenskiöld and Palander.[102] I had no time at all, especially since my mother lives in the country, not in the city.

Yesterday I met Major Wennerström at our establishment.[103] He told me that Mrs Cederström's daughter[104] is near death and will likely die within the next few days. When I think of her strong, healthy constitution, but the doctor, they say, ruined her stomach with morphine injections, so that she could no longer keep down any food. Liedbeck also saw her recently and said she was like a walking corpse. That shows how people fare who swallow too much medication, and I write this to you as a warning. Now, my dear little child, I'll go into the country to see my dear mother and will stay there all day. Tomorrow I drive to the factory.

Fond greetings,

Your Alfred

---

102 For Nordenskiöld and Palander see Letter 45.
103 Carl Wennerström (1820–1893) co-financed Nobel's Swedish Nitroglycerin Company in 1864 and became one of its directors. The factory in Vinterviken near Stockholm was built and equipped with the assistance of Nobel's old friend Alarik Liedbeck. See Letter 3.
104 Unidentified.

65

Brussels, 11 September 1882

My dear, sweet Sofferl,

I sent you a telegram today from here where I am resting a little en route to Amsterdam. I want to consult Metzger[105] there concerning my migraine. Perhaps something can be done.

As you may have guessed, I wanted to use the baths in Aachen. I reasoned like this: the water in Aachen is like that in Aix-les-Bains, containing sulfur, and in much stronger concentration. I could therefore save myself a long journey if I could complete a treatment in Aachen. But my calculation was not correct. I took only three showers in Aachen and developed such a profound migraine that it hasn't improved even now, and I can hardly write these lines. I have no explanation for it since I have not caught a cold, and it must be purely the effect of the baths.

Here I am, writing almost two pages about myself alone, but that is only because I know that you take a kind interest in how I'm doing.

I have received your nice telegram and am glad to see from it that you have greater hopes of the treatment in Franzensbad doing you good. But you write that you want to leave Franzensbad already on the 14th or 15th. That cannot be. It would cancel all the beneficial effects of the treatment because your menses will come on the 13th or 14th, and on those days you absolutely cannot travel. For heaven's sake, or rather for your own sake, be reasonable and don't commit a folly that may affect you later and throughout your life.

Don't forget to tell the physician in good time when you expect your menses so that there is no carelessness with your bathing.

How is little Olga? Give her my regards. And is Bella[106] alright again? You can imagine, all that travelling at night has made me tired and distracted, so that I forgot my coat in Helsingör and had to send a telegraph for it.

You ask what you are supposed to do about Marie. It is difficult to give you advice. In my eyes she has great faults and few merits. I see

105 Johann Georg Metzger (1838–1909), a physician and well-known masseur. For his advice to Nobel see Letter 66.
106 Sofie's dog.

only one reason to take her back, namely that you have no one else at hand in Paris. But she is a heartless, unpleasant, and capricious thing.

And now, child of my heart, I wish you as many sunny days in your life as there were rainy days in August, that is, every day, and I embrace you fondly in my thoughts.

Your Alfred

66

Grand Hotel d'Aix-les-Bains [21 September 1882]

My dear, good Sofiecherl,

I just received your telegram from Wiesbaden, from which I see that you arrived there somewhat ailing. All that gallivanting around in the world is not good for you, dear little child. You need to stay in one place where they take good care of you, and go to Franzensbad only once a year at the most. All other therapies are hardly advisable in your condition.

I too arrived sick once again. I have a running nose and a cough, which I caught in Paris. As you know, those ailments right away turn into bronchitis and then I will labour over it for months. The physician advised me not to leave the house for now and not to take any baths until my cold has somewhat abated. You ask why I don't want you to come here. The reason is very simple. Mr Schaw[107] and other gentlemen from Glasgow are here and, as you know, all Scotsmen are puritans. I haven't forgotten what insults poor Downie[108] had to suffer in a similar and forgivable situation, and I don't want to make myself the target of such talk. A man can have many pieces of gold or obtain them, but he has only one reputation, which he has a duty to keep as spotless as possible.

There is absolutely no way you can take Marie back. I took it upon myself to write in this sense to Mrs Berzel. Consider the time when you reproached her for spilling water on the floor (in the bathroom), and she had the nerve to say such nasty things about you that even the

---

107 A.S. Schaw, chairman of the board of Nobel's Explosives Company, Glasgow.
108 John Downie, general manager in Glasgow.

other people living in the house instructed their servants not to salute you. From this you can see what a "pearl" you had in her, and of what deceit she was capable.

I paid the insurance fee for you in Paris. It had already expired on 2 July. The baths in Aachen appear to be no good for me. Already on the second day I developed such an insufferable migraine, which got worse after the third showering, and continued for more than a week afterwards. Altogether I took only three showers, and since they were so harmful to me, I saw no reason to continue.

I went to Amsterdam only to consult with the famous masseur Dr. Metzger. You know what he said to me? "You have a chronic inflammation of the muscles, but not just locally. Almost all your muscles are affected. This kind of illness is my specialty, but I can't help you because it is a genetic illness, and all my skill would be useless. You must try Marienbad or Franzensbad, but a year isn't enough to cure your illness. You may not even feel an improvement the first year, but you must continue. In the end the therapy will help you. This year it is too late to go there. You can take a few baths in Aix-les-Bains, but the Bohemian baths will be more useful to you."

What do you say to that? I suppose I will stay here until the end of the month. Write a few lines, dear child of my heart. A thousand heartfelt greetings.

Your old friend Alfred.

I would have written long ago, but I was uncertain how long you would stay in Franzensbad, and I had no other address for you.

67

London, Army and Navy Hotel, 22 May 1883

Little child of my heart,

What precious weather we are having here and what a pity that I cannot enjoy it in the country.

I understand that Carlsbad is full of people, and one probably has to make reservations for an apartment two weeks in advance. Robert wrote that he will go to Marienbad the next time and undergo therapy there. He must be a really healthy fellow to survive all those cures – he has the condition of a giant.

Do you know whom I had as a companion in the train compartment? Archbishop Croke,[109] who is by the way a well-informed and pleasant man.

The crossing was wonderful. Not a breeze and no waves. Much more quiet than Paris. Tomorrow morning I'm off to Scotland, but I won't stay long. Prepare what you can prepare for your departure, but don't fatigue yourself at the dressmakers'. It is nonsense to make yourself ill on account of some rags. Don't get anxious. I'll be back soon. Don't worry because it harms the digestion and don't get angry because that is quite useless.

In short, be all through a reasonable and good child. With this fond wish I send you a heartfelt letter-kiss on your sweet mouth.

Your Alfred

Here it was a great bother to obtain a suite. London is overrun because of the great race.[110] I tried at least 23 hotels and finally found quarters in a hotel that had just opened up yesterday.

Write to this address:
Mr A. Nobel
St. Enoch's Hotel,
Glasgow

68

Marienbad, 18 August [1883, envelope stamped in Aix-les-Bains 1 September 1883]

My dear, good Sofie,

Unfortunately I can see that your health is suffering once more. It is quite sad to think that you spent your time in Ischl likely in vain instead of undergoing proper therapy this year. You have certainly harmed yourself with your silly desire to go to Ischl, and it has spoiled my whole summer. I had to drive around like a courier or a salesman and can't even take the time to gladden my old mother with a visit in the summer.

---

109 Thomas Croke (1824–1902), archbishop of Auckland until 1874, then archbishop of Cashel, Ireland.
110 The Oaks Stake or Epsom Oaks, named after the estate of the 12th Earl of Derby, who established the race in 1778.

You must understand, my cute little child, that you should obey me rather than dictate to me. Your wishes, caprices, fantasies, and whims don't add up to anything pleasant.

Try your luck in Carlsbad now and take as many mudbaths as possible in Franzensbad. I see from Lehman's book[111] that iron-vitriol baths, which are more or less the same as mudbaths, are good against intestinal catarrh. I am very sorry for you, my poor child, that we still haven't overcome all those problems. I too am not in very good condition. I have no appetite, I keep drinking Salzquelle and take strong mudbaths, but so far I see no success and suffer a great deal from migraine and feel terrible. Perhaps it doesn't suit me to live so completely on my own as I do, while undergoing this therapy. Since my arrival here I haven't exchanged a single word with anyone, and so I get more and more used to brooding. By the way, I am overwhelmed with work here. The letters mount up terribly. Today I sent off eighteen already. Yours is the nineteenth. All that letter-writing does not go together with a therapy. But what can I do? I can't leave those letters without reply, and to find a secretary for my correspondence is not easy. It is a difficult task to write letters with technical content in five languages. You ask what you should give the young doctor and the concierge. Give Steiner something like 20 to 25 florins for five consultations. As far as the concierge is concerned, I don't know what is usual there, but ask Geschwandtner or Gottwald,[112] and be liberal with your pay. Don't forget any bills that are due and keep everything in very good order. But don't start to pack or to do anything fatiguing until you feel a little stronger. Should I reserve a room for you in Carlsbad or in Franzensbad or what are your plans? Ludvig will remain another two weeks in Ems, and will find it difficult to go there afterwards. But you shouldn't relinquish your right to move into the apartment in September, or what do you say? A thousand fond greetings and embraces.

Your Alfred

[Addressed to Madame Sofie Nobel, Hotel Pupp, Carlsbad, Bohemia]

---

111 Johann Gottlob Lehman (1719–1769), author of the metallurgical handbook *Abhandlung von den Metall-Müttern* ... (1753)

112 Businessmen in Ischl. Georg Gschwandtner is listed together with Heinrich Gottwald as a member of a local club, the Alpenverein. He was the cashier in 1874 (*Zeitschrift des deutschen Alpenvereins*, 1874). Heinrich Gottwald chaired the Ischl association of merchants, 1876–1880 (*Statistischer Bericht über die gesammten wirthschaftlichen Entwicklungen des Kronlandes Salzburg* [Salzburg, 1883], p. 141).

69

Dresden, 4 September [1883]

My dear Sofiecherl,

You probably got my short telegram yesterday very late. The train was much delayed so that we arrived only around 7:30.

During the journey I had an opportunity to observe how appearances can deceive. You too would have noted the undistinguished features of my fellow-travellers in the compartment. I liked them so little that I did not speak a word to them for some hours. Then I noted in the youngest gentleman a refined smile that hinted at education. So I began a conversation with him, and soon we were deeply involved in a scientific discussion. He was a Dr. Thomson from London.[113] He knew most of my acquaintances there and is himself quite an interesting character. Time passed so quickly in the meantime, that I was quite surprised when we arrived in Dresden with a delay. There your dear little telegram was waiting for me, and another one from Ludvig, in which he announced his arrival tonight. So far I have had no message from Mr Schaw,[114] and so I sent him another telegram. This sudden demand for money in Scotland was inconvenient for me because according to current views about Russia, I will have every reason to take payment in stocks rather than cash.[115]

Write immediately. I am very curious to hear how your treatment works and if the sparkling water is really useful. To improve your appetite, I advise you to keep driving to the Kaiserpark – even the name should attract your loyal heart – and walk and sit with Olga and Bella in the spruce forest. Unfortunately the dog does not set a good moral example for Olga. If you hear of a convent for dogs, my advice is to make her take the veil, because that is the usual remedy for unrequited love.

Godspeed and think sometimes of your fond lover.

Alfred

---

113 Unidentified.
114 For Schaw see Letter 66.
115 On the difficulties in Russia see Letter 72.

70

Dresden, 7 September [1883]

My dear, good Sofie,

Since my digestion has worsened in consequence of the strain, I did not dare travel through the night, and so I stayed until this morning. My train starts at 8:25 here and I will, if possible, travel through to Brussels, that is, practically the whole night.

Ludvig looks much better than the last time I saw him. That's because he has few worries and is burdened with less work. Crusell[116] is here as well, and a colonel whom my brother will send to Baku – a short, negligible journey of eight days (and nights) without taking a break!

It is strange: The representative[117] of whom I told you, the man who bought a great palace here, lives there all by himself with his unprepossessing wife. The good fellow, who looks otherwise like a staid bourgeois, is making himself ridiculous. He would like to palm off his castle on Ludvig, but he would not be tempted.

Ludvig asked for your address. He wanted to send you something, I think.

I notice that your health always deteriorates in the first days after my departure. But as soon as you make a new acquaintance you are no longer bored and your mood turns sunny, your temperament lively, and the digestion in your disordered cooking-machine rallies. I assume you have wonderful weather there. Enjoy the air in the forest, be happy like a little bird, flirt moderately, show tact and reason, and I send you a thousand fond greetings and embraces.

Your devoted and loving Alfred

---

116 Hjalmar Crusell, Ludvig Nobel's illegitimate son who was the head of the laboratory in St. Petersburg.
117 *Agent* in German; unidentified.

71

Ostende, 9 September [1883]

Dear love,

Since I am staying here, waiting for the steamer that departs for Dover in the evening, I sent you a telegram asking you to address a message to me here and tell me about your health. So far I have received no telegram and I fear I will have to depart without receiving the calming number 18.[118]

The coast is splendid and I bet that a summer spent here would have been more beneficial than Ischl and Carlsbad together. But you, my child, have your own will, which unfortunately never puts together anything reasonable. Here there are lots of villas for rent with kitchens, so that one can arrange things very comfortably. I know of course that nervous people at first suffer neuralgia from the sea air, but after a short while the pain disappears. Many people can't take bathing in the open sea, and that is indeed not necessary for getting stronger.

I am very curious to hear whether Carlsbad really brings you relief. I hope so, but around the 14th or 15th you must be very careful. Don't eat anything that could harm you and guard against catching cold as if against the plague.

The night train from Cologne to Brussels fatigued me a great deal, I fear that I have already lost any benefit that my stay in Marienbad may have produced. But what can I do? As long as I can't get rid of that cursed responsibility, my directorship, I have no peace. Since time is very short, I have decided to travel the five hours from here to Dover at night. I hope to be able to sleep and to arrive in London without feeling too unwell.

Again, my dear Sofiechen, let me warn you. Be on guard in every way and don't think if you feel better one day that you can rush off. A thousand fond and heartfelt greetings.

Your Alfred

---

118 Either a reference to the number of the letter (they numbered their letters to keep track of them as Nobel travelled and the letters had to be forwarded; see Letter 73) or to a code they used to keep certain matters private, especially in telegrams, which were read by the transmitters. See Letter 86.

72

21 September 1883

My dear, sweet little child,

My heart is sore when I think that you must roam the world all by yourself. But it's your own fault. I have been telling you for years now that you absolutely must have a chaperone. And almost everything you have suffered and all the unbearable things I have born are the consequence of your not obeying me in that respect.

Can't you understand what a terrible burden it is for a busy man like me to have a companion who knows absolutely no one and therefore keeps me from moving freely in society? This is a terrible state of affairs that, over a short time, has aged me twenty years. Whereas if you had a chaperone – a very decent and reliable person of course – I wouldn't have to travel all over Europe almost as if I were your nurse. And on the whole we would spend more time together and lead a more tolerable life. Believe me, your caprices and silly childishness are a source of trouble. It is not so difficult to please me, and although I am often in a bad mood, I'm not quarrelsome.

Today I sent you a long, calming telegram. I ask you earnestly not to get yourself into a snit after that strenuous treatment.

We might talk about you travelling to Montreux[119] and spend the winter there. In Paris at any rate the weather is not yet cold and won't be cold for some time. Don't you find it rather droll that you have an apartment here? You are away all summer, and in the winter you don't want to come here either. That's quite alright in my opinion. You are not suited to Paris, and Paris is not suited to you. But why don't you choose to live where you really want to settle down? Montreux or wherever else you please. But to gallivant around as you do, and to make me rush around when I am so busy that life has become bitter like gall to me – that is neither wise nor fair. Don't you think I have enough obligatory travel without the endless travel you impose on me?

---

119 On Lake Geneva, Switzerland

The disorderly state of affairs in Petersburg[120] is unprecedented and makes my hair stand on end. They rejected two bills of exchange for Gen. N.[121] No, not because there was no money but because they had been sent to a bank other than they wanted. What do you say to that? But don't tell anyone about it and burn this letter at once.

Ludvig's address is simply "Mr Ludvig Nobel, St. Petersburg." Write in the lower corner of the envelope "personal," or they'll open the letter at the office.

With fond greetings and many heartfelt kisses, I remain

Your loving and devoted Alfred.

73

August [1883?]

My dear, sweet Sofferl,

You can see how difficult it is to find a free moment. I had to depart from Vienna without driving to […][122]for a fitting of the clothes I ordered. I started my letter to you three times and got only half-way, without being able to finish. Max[123] is a very pleasant man, but he doesn't understand that he is sometimes a nuisance. Luckily this won't go on forever. It is 5 a.m. as I am writing these lines, because later on I can't even breathe without witnesses. Your letter concerning the samovar arrived too late. It wasn't possible for me to obtain one. I ordered twelve bottles of Tokay by telegram. The pills and prescription you will receive from the apothecary directly.

Max's visit, although it meant more stress, was quite welcome. He was able to see and take note of quite a bit, and his journey may lead to a connection between the two companies, which would be very desirable.

---

120 Ludvig Nobel was the driving force behind Branobel, the oil company founded by the Nobel brothers in 1879. The 1880s were difficult years for the company – two oil tankers were seriously damaged, oil prices fluctuated, and Standard Oil became a competitor in the world market. See Letter 54. On the difficulties in the years 1882 and 1883, see Tolf, *The Russian Rockefellers*, pp. 80–3.
121 Unidentified. The reading is uncertain.
122 Illegible.
123 Max Philipp, director of the German Society of Dynamite. He acted as one of Nobel's lawyers in Vienna. After Nobel's death, he bought his villa in San Remo.

I had news from Glasgow. On my initiative the directors there wanted to put restraints on Cuthbert,[124] whereupon he put in his resignation. Notice must be given six months in advance. He will leave after that period.

My health is not good. Dining with the gentlemen instead of keeping to a diet after the Carlsbad spa has greatly affected my digestion. Therefore I'd rather do without the long journey to Aix, which would take more than three days. But I am uncertain how to proceed. Will the mudbaths in Marienbad help me or not? I am up against the unknown, whereas Aix helped me without a doubt. I would very much like to ask Borges[125] in Marienbad to tell me in all sinceritywhether he has some assurance that the treatment there will help me and will not be a complete waste of my time. After that I'll make my decision. In any case I need one or two days of rest. I have to have my shirts laundered and wait for my letters that are in transit, but I can do that very well in Marienbad, since it's on the way to Aix.

The Paris letters #14 and #15 finally arrived in Vienna, but with a delay. #13 is still missing, and that is very unpleasant because one can't know whether there is anything important in it.

Take good care of yourself, my dear, good little child, and be on guard especially in the days from the 18th to the 20th. Don't walk too much, don't eat food that is too heavy, don't catch a cold, and don't get angry, I beg you, with a thousand heartfelt greetings and kisses.

Your Alfred

74

Aix-les-Bains, 17 June [1884]

My dear, good Sofiecherl,

An hour ago I received your nice little letter and thank you heartily for it. Since my arrival here I am beset by work. The people in Glasgow don't leave me in peace and with this endless writing it is impossible

---

124 Alexander Cuthbert, manager of Nobel's company in Glasgow.
125 A Dr "S. Porges" appears repeatedly in literature on Marienbad in the 1860s. He is the author of *The Rudolfsquelle in Marienbad* (Berlin, 1868).

for any treatment to take effect. Today I wrote a letter to Glasgow that was seventeen pages long and I feel light-headed as a result. But those people are in such difficulties that no one can help them in the long run – neither I nor they themselves.

Forgive me if I write only a few lines. The treatment would certainly make me feel better if it weren't for the work. The showers I took have laid me low and given me a lasting fever. I spent most of the journey reading Turgenev's novel,[126] which I brought along. It is very nice, natural, and moving. I will send it to you and am sure it will shorten your time, if you have any leisurely hours.

I hope that the treatment will truly benefit you this time, my dear child. But this also requires a careful diet, very little wine and no Chartreuse at all. Drinking too much sparkling water can also be bad.

Give my regards to little Olga.

Heartfelt kisses,

Your Alfred

75

[Aix-les-Bains, 19 June 1884]

Child of my heart,

If they say I am staying at the hotel, they are using the wrong word. It is a hospital, with the requirement that the least sick people must have had at least three strokes. If anyone takes more than a step per minute, he is a relatively healthy person. But that's not the worst. Looking at those terrible eczemas, pimples, abscesses, and other skin problems on their faces brings on a wave of something like seasickness, which can only be calmed by wearing very dark glasses. Schiller's[127] verse keeps

---

126 Ivan Turgenev (1818–83). The book is probably *Das adelige Nest* (Hamburg, 1884), a German translation of *Dvorjanskoe gnezdo*. Both the Russian original and the German translation were in Nobel's private library. See the list at http://www .nobelprize.org/alfred_nobel/library/fiction-ru.html.

127 Actually a biblical notion, found in, for example, Ps 84:6. Friedrich Schiller was one of the foremost German dramatists (1759–1805).

coming to mind: "The world is a vale of tears." To recover from all this misery I thought I'd go to the theatre. *La Boule*[128] was on the program, a nice comedy. But when the actors and actresses laughed, they showed teeth so rotten that an odour as from a corpse pervaded the theatre. Their faces, too, reminded me very much of an anatomical museum, and on my return I found those rotting half-corpses in the hotel almost pleasant. There was one fifty-year-old woman – a paralysed hunchback, toothless, full of zits, with a red nose and a mustache, pock-marked, with sweaty feet, dirty fingernails, and a stutter, a woman who washes once a month, slovenly, sour-smelling, red-eyed, flat-footed, dressed like a German, but greasy like a Russian, taking snuff and chewing tobacco, and this beauty thought I was mightily in love with her exquisite charm. That is how deceptive appearances are. It was only a matter of contrast between her and the actresses, who were more unappetizing by 6%, as you put it.[129]

What blessed company for someone like me who is, since my arrival, overwhelmed by work! I am so burdened with letters and problems pressing down on me, most of them coming from England, that I must joke about them, or my eyes would deliver material for saltwater baths.

But whatever one may say, the surroundings here are beautiful, the establishment very nice, the food excellent and not expensive, and that's not to be sniffed at by a man like me, who loses a million francs every month. I must pay attention to that. And I could save even more, for the moment I regard my fellow men, I lose my appetite.

It is a little cold here, but the weather is beautiful. Unfortunately I have so much work that I can hardly leave the "hospital."

The spa is good for me, but not to exchange a word with anyone for days makes for a bad mood after all. Don't worry about the apartment. Stay where you are with or without shutters. No one in Carlsbad looks into windows. They look only at the doors that bear the blessing of Carlsbad and the sign "Here." The charms of women are ignored, and that's good for female decency.

I wanted to make you laugh – I don't know whether I succeeded. I am sealing my joke with a heartfelt kiss – "here."

Your Alfred

---

128 *La Boule* was one of the many farces and light comedies written by Henri Meilhac and Ludovic Halévy in the 1870s and 1880s.
129 A jocular turn of phrase, meaning "by very little."

76

[Hamburg, 11 June 1884]

My dear, little Sofferl,

From your telegrams I see that your condition after the treatment in Carlsbad is not the best. I expected that. In my opinion true peace and good care is the best treatment. You insist on seeking it in faraway regions, where I cannot and will not follow you. That is how it has been now for almost seven years, and it is going on in the most senseless manner, without any benefit to you and at great sacrifice to me, which embitters and takes a toll on my life. I want to devote all of my time to my profession and to science, and consider women generally – young or old – as bothersome interruptions that rob me of time. I've been singing that same song to you for seven years now in every key without you being able or willing to understand me. And instead of doing my laboratory work, I act like a nurse for an adult child, who believes that she is at liberty to impose on me all sorts of whims, caprices, and fancies. If you had been satisfied with a pleasant and healthy place in the country where I live, you could have spent your days in happiness and would not be an endless bother to me and, in addition, make me a laughing stock among all my acquaintances. Unfortunately we share a past that is full of bitterness for me. I would forget it all, but lost time cannot be recovered, and that thought leaves me restless day and night.

But let us not talk about what is past. I only preach to you in the hope of opening your eyes a little. What can be done in future? You want a villa in Ischl. Alright: we'll buy it, and then? I am no more willing to visit Ischl in future than I am to visit hell. Olga will have to deal with her studies, and cannot keep you company, at least not for the next year. In the circumstances you will not like Ischl, and then you will sing the old song of complaint, wanting to stay in a villa in Reichenau or Villach, or Görtz or Arco, or Mürzzuschlag, or what do I know! One complaint after another, and why? Because you don't want to understand that your position is wrong and that a little courage is needed on your part to improve it. You could do it very well, and I will gladly, indeed with my whole heart, help you. But you must stop the mad gallivanting around which you have done so far and stop bothering me all the time with your ideas, when I already have to travel so much. I think you don't notice how unprecedented it is to claim that there is no place in this large country where you can live and to insist that I, who am your protector out of pure good will, should give in to your childish whims.

I hope that you will have completely recovered from the treatment in Carlsbad at the end of this month, and that it has not been without benefit. As far as I am concerned, I am so tortured and suffer so much from stomach ailments and headaches that you must not be surprised if I don't write in a more cheerful tone.

I enclose an English bill in the value of 50 pounds, my dear little child, which is worth about 600 florins. That's all I have with me and I hope that it will last you for a while. Let me know.

I am so busy during the day and far into the night that I can hardly send you a telegram. I am writing these lines at 2:30 in the morning. That is how long the meeting lasted.

A thousand heartfelt greetings and embraces from your sympathetic old friend,

Alfred

Hamburg, Thursday night

77

19 June [1884]

Little child of my heart,

I have in front of me your letter of 15th June and see with some worry that the waters of Carlsbad still haven't benefited you a great deal. I see I must give Olga instructions to guard you well and make sure you make no mistakes in your diet.

It doesn't take a lot of brains to tell you that this so-called Excellency is a stupid womanizer. Let him go because people who start like that are not worth dealing with. I have never heard of a Darwin other than the man who looks remarkably like an ape himself and has attempted to show that human beings are descended from apes.[130]

Yesterday I received a very kind letter from a real Excellency. He could not have been more obliging, for the finance minister sent me a

---

130 I.e., Charles Darwin (1802–1882).

decree signed by the president during my absence in London, which appoints me to a higher rank in the Legion of Honour.[131] I don't value the appointment in itself much, but the manner in which I was informed of it moved me. I begin to notice that the French can actually be kind.

I paid your rent and enclose the receipt. Don't lose it.

From your short letter I see that Marie has joined you again after all. I suppose you don't have the courage to hurt anyone.

Together with this letter, you will receive a little memento of a certain little boy, which I send you on his request.

I must explain why I am still sitting in Paris. Once again we had two accidents, one in Paulilles and another in Portugal, which give us a lot of trouble.[132] The factory will be closed down, and of course we must fight against that. We had a meeting about it on Friday, and I had to make it known that I would be present at it.

Do you remember the address of Miss Martinetz?[133] She has to give me the date of the receipt, or I won't know when I have to pay her the interest.

Give my regards to my brother when you get together with him and let me embrace and kiss you from my soul,

Your Alfred.

Do not trust this so-called Excellency. He must be a crook. Is he not a Russian?

77a

Telegram #1216, Aix-les-Bains, [June 1884]

Will have to break off treatment soon. Would like St. Moritz but fear business affairs will keep me away. More tomorrow, telegraphic kiss. Alfred

131 Nobel received a number of honours. He was a member of the Royal Swedish Academy and the Royal Society in London, and a Knight of the Order of the Polar North.

132 Nobel began manufacturing dynamite in Paulilles (southern France) in 1870, when the Franco-Prussian War broke out. His Portuguese factory (established in 1873) was in Trafaria, near Lisbon.

133 Unidentified

78

Vienna, 9 August, 6 a.m. [1884]

Little child of my heart,

I received your telegram of yesterday just in time to miss the evening train to Berlin. I couldn't leave your telegram unanswered of course, and so I had to countermand my orders to the porter concerning my departure.

Although your request is extremely unreasonable, I would have granted it if at all possible. But Dr. Scharlach[134] has been informed by telegram that I would arrive in Hamburg on the 10th at the latest and would travel on to Stockholm on the 11th in the evening at the latest, and would return to Hamburg around the 20th. Scharlach has already left [Brennerbad], and I have no address where a telegram could reach him. Therefore it is impossible to think of any change of plan. I would not want to face the embarrassment.

I admire all the nonsense a little woman can develop in her little head! No idea is mad enough to fail to flourish there. You want to drive back and forth between Ischl and Vienna to spend a few hours with your nasty old grouchy-bear. In any case the journey would be too risky, because you would certainly get your menses during the trip.

End of sermon, because the trains don't wait and the hour of my departure is near.

Best wishes,

Your Alfred

I would have departed yesterday morning already if it had not been necessary to telegraph back and forth with Scharlach and I was delayed by Dr Smith with the business of the patent.[135]

---

134 Julius Scharlach (1842–1908), lawyer at Nobel's company in Hamburg.
135 Perhaps H. Julius Smith, who patented a detonator in 1868.

78a[136]

Paris, 26 August [1884]

My dear, good Sofiechen,

Since my return here there is such a rush – you can't imagine it. It has made me so nervous that I can barely hold the pen. Mr Roux has left the French company, so to speak, and has been replaced by Barbe as director of the business.[137] This is now being carried on so energetically that we have meetings lasting four or five hours every day. It is pure torture, and yet I cannot stay away since Mr Barbe's predecessor made a pretty mess of everything. In addition a mountain of correspondence has built up. Individual projects of the company make a great deal of work for me, and then there is the most interesting work in the laboratories, which cannot progress without my direction. All of this makes me sick with nervousness.

As long as I am travelling, it is only the correspondence that weighs on me, but as soon as I come back here, there is so much else to do that I cannot find enough time in spite of the greatest efforts.

In Stockholm I found my mother in better health than I had dared to hope. If she continues to improve, the dear old woman will live another twenty years. I also visited my brother Robert. His mood has improved a little, but he is still the same hypochondriac. Every day and every hour he imagines a new illness. In the morning it is a heart murmur, at breakfast gout, in the afternoon tuberculosis, in the evening blood poisoning, and at bedtime cancer, which attacks either his tongue, or his kidneys, or his chest. I am not exaggerating. It is literally true. Now he wants to go to Carlsbad, and after that he doesn't know where he will spend the winter in a suitable climate.

Your letters make me worry about your state of health. But it makes no sense to gallivant around in the world when one isn't feeling well. What are you looking for in your many travels? Reason perhaps? No effort will obtain that for you, and yet you dearly need it. Sometimes I think Bella acts more reasonably than you. You can be animated, amiable, and even witty, but never reasonable. What you lack first and foremost – unfortunately – is an understanding for the feelings and

---

136 Number 35 in Arkiv ÖI-5.
137 Louis Roux (1823–1905), explosives specialist and chair of the board at Nobel's French company from 1875. He had been the manager for black powder production for the French government. For Barbe see Letter 18 and note.

aspirations of others. And that is the whole secret of winning a man's heart. Everyone can be self-centred, but to sense when one hurts other people's feelings and how to avoid it – that is the characteristic of a true woman. For that you need an innate sensibility, which you have to some extent, and culture, which you completely lack. I write this to you so that you may think about it for a while and improve in that respect.

You write that loneliness has a detrimental effect on your health. But that comes mainly from your total lack of occupation. Such ongoing idleness is unhealthy for anyone. You neither work, nor write, nor read, nor think. How can you not be lonely in these circumstances?

Now there is space only to send you a heartfelt word and a thousand fond greetings,

Your Alfred

79

Paris, 29 August [1884]

My dearest little child,

I see with great joy from your last telegram that your troubles have eased somewhat in Vienna. I keep telling you that Ischl isn't suitable for you – neither the cool weather, nor the dampness, nor the bad water. But you are suffering from an inflexibility, which actually deserves a better brain and a firmer character. For if one lacks reason oneself one must accept the advice of others. But you accept advice only immediately after you have committed a great stupidity or carelessness and even then you obey only as long as you personally suffer from the consequences and are bothered by them. All children are like that, and it appears one can't expect more from adult babies than from the little ones. Indeed, when they suffer from diarrhea they very much resemble their miniature doubles, because those little ones are just as helpless in that respect.

Here it has been cool for the last few days and autumn-like. I suppose in Vienna the weather will have cooled down as well so that you won't have a reason to look for cool air on the Semmering.[138] Rather, guard against catching a cold and wear woolen underclothes. Don't get the idea of staying in mountain villages because September is always cool

---

138 Mountainous area south of Vienna.

there. Don't eat fruit, especially not forbidden fruit. Look after Bella and permit her a black-and-golden husband, but one that treats her gently and allows his noble wife to bite him without growling.

During dinner I allow you to complain a little, since your digestion doesn't work without it. Ill humour must out, or it will spoil your stomach. Unfortunately another wave of ill humour will succeed immediately. This is called female kindness and pleasantry. To be consistent, let's also call gall "sugar."

Write diligently how you are, save your health, don't pamper driver 665,[139] who has already been so greatly pampered by you with tips, don't cry when there is a thunderstorm, don't tremble before the devil, honour the emperor, admire the empress, pray for the well-being of both, greet little Olga although she is not of such noble birth, throw stones for Bella, and accept my heartfelt embraces and kisses.

Your Alfred

80

[1884]

My dear Sofiecherl,

Although you make holy promises, especially on the first page of the letter I just received, or rather because you make them, I can see that you amuse yourself well without me and can very well do without my presence. In your previous letter you even write that I need not take the trouble of making unnecessary journeys, from which I conclude that you'd prefer not to be bothered. Don't worry, I certainly won't bother anyone. Nor should you write such long letters. Your style tells me that you blacken all that paper only out of regard for me. But young ladies who need money from old gentlemen consider it their duty to pen long compliments and bow respectfully, and all the time they think of others, and yet regard one as stupid enough not to see through your tricks. That would be all very well if you didn't use my name everywhere so that people will laugh behind my back on every street corner. In your long letters I miss what is most important, namely whether Bella has obtained a brown-haired or a blond husband, and whether many celebrities attended the wedding.

---

139 Code? See Letters 71, 86.

What were the bridesmaids like when Bella said "yes"? Did they blush a lot, did they hide their faces behind their paws?

I sent you two registered letters, the first one with 400 florins, the second with 2,500 francs, but I am not pleased with your bottomless spending and find it is high time you stopped your nomadic life, especially because it is an embarrassment to me. I spend almost all my time at Sévran and have completed more work in the laboratory there in one week than in three months when you are here.

I have never seen Paris so empty of life. Anybody who could, departed and is gone, and there are no foreigners either. Fear of the cholera is the reason for it. All restaurants and hotels are empty. I dined day before yesterday at the Grand Hotel – there were eight persons there!!! Normally several hundred people dine there at this time of the year.

The weather has gotten worse, and ever since I suffer once again from headaches and rheumatism. Nevertheless I'd like to do without the journey to Aix and rather spend time usefully in Sévran. I therefore drink the mineral water of Aix here and drive out daily to Enghien to take a sulfur bath there.

My driver quit. He didn't like it when I criticized him for being half an hour late with hitching up the horses when I ordered them for a certain hour. So now I have the problem of finding a new driver and can't go anywhere until I have one.

I am very glad to hear that your health is improving and that your days are amusing. Well, your entertainment doesn't come cheap, which doesn't matter to you, but the greatest cost is to my honour and name, which is being abused in that manner. You have neither the correct feeling nor the wit to protect the reputation of another person. And yet I am fond of you and would be even fonder if you would seek to marry a good, young, honourable man,[140] and be a good wife and fulfil your purpose in life. Gallivanting around with admirers, you will end up on a slippery slope and for that, I am sure, you are too decent, too good and gentle. I am sorry that you don't understand that, but your brain is good for nothing.

Give my regards to Olga.

A thousand heartfelt greetings and embraces,

Your Alfred

---

140 For Nobel's interest in arranging a marriage for Sofie, see a letter from her father to Nobel, Appendix, p. 277. It was not unusual for a wealthy lover to make such arrangements. The practice was immortalized by Dostoevsky in *The Idiot*.

81

20 September [1884]

My dear little dove,

I would have written long ago, but I had no idea that you would stay on in Ischl and did not know whether a letter would reach you there. In any case I have a hard time understanding why you remain in Ischl for so long when you are in a hurry to get to Merano because of the accommodation. The only explanation I can think of is that you have found someone with whom you can pass the time and who gives you a new opportunity to embarrass me. Really, your travelling around in the world under my name, either alone or with your little companion, is rather strange. If you had the least feeling for me or gratitude, you would understand without any explanations, how badly you hurt me and how unfair you are.

Fehrenbach[141] has long ago left for Glasgow, and I am waiting for a message from him whether it is absolutely necessary for me to travel there. The weather is said to be awful there, and I would like to avoid the journey especially since my neck is almost paralysed with rheumatism.

On my return journey, it was ice-cold until we reached Strasbourg. From there on we experienced a brooding heat as if it was summer, so I suffered torment in my winter suit.

I have followed your instructions to the best of my ability. It wasn't clear to me from your scribbling what it is you want. "Pieds" can only mean "feet," and they don't sell feet at Louvre's, so it was not possible to send any to you. Cut-off human feet are not to be found among merchandise and articles of fashion in civilized countries. I ordered a dress at Moret's – in a blue that is a little lighter than navy. But Moret doesn't have much that is pretty, and I wouldn't recommend buying much there. The coats, too, are all too gaudy. I couldn't tell them where to send the dress, given your vagabond life. At Louvre's I bought gloves, ruffles, veils, and cravats in great quantity, which I sent to Meissl's Hotel, Vienna.[142] These things are supposed to arrive there on Wednesday. I

141　For Fehrenbach see Letter 11.
142　A legendary hotel in the centre of Vienna, at 2 Neuer Markt, where people stayed "who don't want to be seen" (Felix Czeike, *Historisches Lexicon Wien*, vol. 4 [Vienna, 1995], p. 236).

couldn't buy any hats at Reboux because I didn't know where to send them. Nor could I order anything for Olga because Louvre's doesn't have ready-made dresses in all sizes, and then there is the problem of not knowing where to send them. All of this was a great nuisance to me because I had to go away and couldn't very well delay my departure on account of a few rags.

Here I live in this metropolis all by myself, separated from all people, so that my life often seems empty and sad to me. At my age every man has the need to have someone around for whom he lives and of whom he can grow fond. It was up to you to be that person, but you did everything imaginable on your part to make such a relationship impossible. Right from the first day I told you: Get the necessary education, for it is impossible for me truly to love a person in my heart, whose lack of tact and culture embarrasses me daily and indeed hourly. You don't feel this lack or you would have done something to alleviate the problem at least partially. Even if I were head over heels in love with you, the kind of letter you write would have the effect of a cold shower. One automatically thinks of the shame and embarrassment resulting from a woman writing – no, scribbling – in this manner under another man's name and sending these nauseating epistles out into the world. Believe me, my dear, good little child, if a person has no understanding for culture, she is suited only for a lower position in life and can only be happy and satisfied in that position. You always believe I can't be fond of anyone, but there you are wrong, I could be fond of you if your lack of culture didn't continuously hurt me. Let me give you an example: A woman is exceptionally fond of her husband, who has decided, however, to step on the corns of her toes every quarter of an hour. Do you think her love would endure? Now, a man's honour is a thousand times more sensitive (at least in my case) than the worst corns, and because you cannot understand that, there have been and still are constant tensions in our relationship. But why do I bother to give you explanations? You don't understand what it means for a person to have honour and pride, or that point would have been clear long ago.

I wish you all the best and pleasant, indeed better days than the days I spend here so very sad and lonely. In my thoughts and in my heart I embrace you,

Your Alfred

82

9 October 1884

Dear Sofferl,

I must expedite the money if I am to send it to you today and have time only to send you and Olga heartfelt greetings. By the way, it seems my greetings leave you quite cold.

    Fondly,

Your Alfred

Enclosed 2,000 francs.

83

Paris, 11 October [1884]

Dear Sofie,

I enclose the remaining 1,000. You will have received the first 2,000. Please return to me the cheque over 2,000, which you could not cash, but cross out my signature on it. I am surprised they did not take the cheque.[143]

    Here I have a great deal of work, and you will pardon me for writing rather succinctly. By the way, the manner in which you said farewell to me gave me the impression that you believe I should be very happy to be your money-dispenser and obedient servant, about whom you need bother very little.

    Heartfelt greetings,

Your Alfred

84

15 October 1884

My dear Sofie,

Reading between the lines I see that you amuse yourself wonderfully in Vienna. Unfortunately I fear that both you and I will be harmed by your

---

143 A cheque drawn on the Banque Russe (see Letter 84). For another bill of exchange that could not be cashed, see Letter 72.

carelessness, although I am quite innocent in this matter. It is strange that you would introduce yourself to a family under a borrowed name, and if you don't understand that, there is something wrong with your mind. But your health is proportionately better, it appears. My heartiest congratulations. I tremble all day when I think that Hoffer[144] is in Vienna now and will perhaps pay you a visit with his vulgar Minchen, which you will have to return out of politeness. For all I know, they'll invite you to go to the theatre – how pleasant for Mrs [...] to see you there in the company of Minchen! What a stew one cooks up when one doesn't have a correct social position, and has no sense for order and not tact enough to keep away from those complex situations and stay as safe as possible. I am heartily sorry for you and would do more if I didn't see that I am losing my lifeblood in this unpleasant story.

My nephew is here and staying with me. In addition I have so much work at hand, mostly because of the meetings that are held here again, that I can't write at length to you.

I enclose some reinforcement for your account, so as not to see you embarrassed. But I must point out to you that I have lost huge sums lately (I will tell you about that in person and don't want it to get around), so that there will soon be a sad end to your great wastefulness.

Return to me the cheque drawn on the Banque Russe and confirm receipt of today's letter. I sent you 2,000 francs, then 1,000, and today again 2,000. I need to know that those letters reached you. Write a line about them.

In former days when I went on a journey, you accompanied me to the train station and wrote to me daily, but things have changed now. I won't complain about it, I merely note that it is so and that my changed behaviour is only the result of your changed behaviour. It appears you wanted to increase the distance between us more and more and at the same time to make ever greater demands on me and use my name more and more. Try to put yourself into my position, my dear child, and you will probably realize how unfair you are to me. Heartfelt kisses,

Your sad Alfred

Today I also have to pay 1,500 francs for your apartment. Make sure to tear up this letter after reading it. I would not want to have that letter fall into the hands of strangers.

---

144 For Hoffer see Letter 11.

85

Paris, 26 October [1884]

My dear, good Sofie,

I write a few lines in great haste. I am dead-tired and tormented by the meetings here, which last every day from 9 a.m. till long into the night. Even today, Saturday, the meetings continued in the same manner, and I must steal a few minutes to reply to you in a few words. It is a terrible life, and I don't see an end to it. At least this time there will be a result, so that all that effort hasn't been in vain.

But now I am so much in need of rest and so ailing that I would rather live in the most miserable village and eat dark bread than suffer this nightmare any longer.

And you, my dear child, are still far away from me. Did you not like Merano – since you want to depart again?

A thousand heartfelt greetings,

Your Alfred

I enclose 700 francs in the hope that this letter reaches you in Munich.

86

Paris, 26 May [1885]

From your long, long letter I see mainly that the Bozen architect would like to make a great deal of money out of this. Believe me, many improvements are needed at the Eilenhof,[145] but it doesn't add up to 7,000.

So you are finally persuaded that the castle is much too remote and very uncomfortable to live in. In any case I am not convinced that Merano will please you in the long run as much as it does now.

---

145 Nobel was prepared to buy a property for Sofie, wherever she chose to settle down. He ended up buying her a villa in Döbling near Vienna (see Letter 139). In spite of what Nobel says about "living together" in an out-of-the-way place, it becomes increasingly clear that he wants to disengage from Sofie, though not without making financial arrangements for her. See the reference to her "new" life in Letter 89 and his hints that she should look for a good husband in Letter 90.

One thing you don't see clearly – that we two can live together only in a place where there are few people. You are so lacking in culture, my dear, good child, that I suffer greatly in your company at the slightest contact with people, and suffer so greatly that life becomes hell for me. You have fine and gentle natural sentiments, but your education was terribly neglected and you have neither the ability nor the desire to catch up on what you have missed. I don't ask for an all-round education, I'm not even partial to that, but I don't want to blush at every word a lady says in her own mother tongue, who pretends to be my wife. Furthermore, your behaviour in public is such that one must ask how such a person obtained such elegant clothes. Worst of all, you don't realize how wrong you are and therefore one can't expect any improvement.

I tell you all of this, not to reproach you, but to deprive you of any illusions about my living with you in a place where one has contact with people. At the first public vulgarity I'll be off and running.

I don't think 26,000 is unreasonable for the Eilenhof and if the owner accepts it, we can sign the contract. Let me know at once when you receive a reply. As for the repairs and furniture, one would have to invest a good amount of money.

Since my return here I suffer once again from severe migraines and a sore throat. The weather is nasty, cold, and windy. Anyone who has a fur coat is wearing it. Last night the temperature was 4 degrees. Since I have hardly any acquaintances any more, my solitary life is very sad and empty for me. It would be different if I could work in my laboratory, but I can't live in Sévran because it is much too damp there. I enclose something to shore up your cash box. Most heartfelt greetings and sweet wishes for my little toad.

Your Alfred

My best regards to Olga.

I'll send a new key to decode my telegrams, so that you won't fill them with silly nonsense that amuses the telegraph ladies in Bozen as well as in Paris.

87

Paris 6 April [1885]

My dear, good Sofiecherl,

All the trouble you take over your villa business! How would you feel if you had as much business weighing you down as I? As far as I am concerned, I am once again so tormented by business that I am almost losing my mind. That is how frazzled I am. The cold bothers me less than the many obligations to write, meet, deal with legal proceedings and technical work – it all ruins my digestion and robs me of sleep. In the first days after my return it is bearable, but then it starts up, and those tormenters of mine are after me from morning until evening.

1,000 francs more or less for a villa where one resides makes little difference, but my silly little child, do you seriously believe that I can live in Merano, which is 1,100 kilometres' distance from Paris? And I bet you won't live there either. Why then stupidly waste money? I am losing enough money in all quarters and am trying in vain to sell my Russian bonds at any price.

I am returning your various building plans. The business in Ischl is too crazy! Now you want to buy a villa in Ischl as well? Why not also in Gmunden, Aussee, Weissenbach, Gries, Reichenau, etc.? And everywhere my name goes back and forth, written on letters and in telegraphs. I am surprised only that the police aren't after me to lock me up in a suitable madhouse.

Eugen Fischhof is here and will marry the daughter of Sedelmeyer on the 9th.[146] Every time I see the man, I like him less. His reputation here is very objectionable, moreover, about which I'll tell you more in person.

I am heartily sorry to see from your little letter that your health has not been the best lately. You can see from this, my dear little toad, that your health does not depend on the place in which you happen to be, and yet you keep visiting places that are most inconvenient for me and where I neither can nor will go, as you know. In short, you lack reason, but your heart is good, am I right?

In your letter I don't see the name of the veterinary in Lugano. Can you give it to me?

Fondest and most heartfelt greetings and embraces,

Your Alfred

---

146 Eugen Fischhof (1853–1926), art dealer in Paris, married Emma Sedelmeyer (b. Vienna, 1862).

88

7 June [1885]

My dear, good Sofiechen,

I am so affected by the water cure and also by the heat that I can only write a few words to you. I just received your short telegram that you sent me after your departure from Innsbruck and am joyful to see that you are much better. The heat in Merano must be terrible, and I don't understand how you could stand it there for so long, or rather, let's say, I do understand it.

With heartfelt greetings and an embrace,

Your Alfred

I enclose only 1,000 francs because a registered letter might present formal problems to you in Basel, but I will immediately send a cheque for you to Paris.

89

Aix-les-Bains 28 July [1885]

My dear little child,

When I wrote to you this morning, I was too fatigued and depressed to say everything I wanted to say to you. I drove out for a few hours and am feeling a bit better now. What I wanted to say to you is this: Be quite honest with me and explain to me how you plan to conduct your new life.[147] I am not asking out of curiosity – that would be a sad kind of curiosity – but to help to arrange things and perhaps save you some trouble.

Again, heartfelt greetings,

Your Alfred, who will soon be forgotten

---

147 The lease on Sofie Hess's apartment in the Avenue d'Eylau was not renewed in 1885.

90

2 February [1886]

My dear Sofferl,

I am so busy that I have no more time to write. I have never before found myself in such a whirl of activity. You ask whether I go out into society. When would I do that? I hardly have time to breathe and none at all to dine, and only rarely to breakfast. And how could it be different, when I spend all my days from early in the morning until late at night in Sévran.

The transfer fee[148] has to be paid, and that must be looked after immediately or all sorts of difficulties and problems may ensue. I enclose the necessary [money]. Complete the transaction as soon as possible. It is past due.

You complain about the Viennese – they are wimps, but not when it comes to demanding pay, and they understand very well how to charge high prices. You will complain about people wherever you go, and you don't see that there is quite a bit to criticize in you, too. Didn't you complain about the French a lot and didn't you treat me badly and without gratitude? On the whole, dear Sofferl, you suffer from great pretentiousness. You consider yourself as so august that people should be proud to ruin themselves for you, to labour for you, and to allow themselves to be maltreated on top of that. Your problem is to find people who approve of your view and perhaps you begin to understand that other people are human too. Your life would be much more pleasant if you could understand that. Your inability to do so is probably a consequence of your terribly bad education. You can be very nice and loving and then spoil the good impression with ill manners no decent person will suffer. You will never obtain good servants, for good servants want to be respected and not be treated like scoundrels. You can find a good husband only among people who will put up with your lack of tact and won't find it a bane. You will always quarrel with people who have a sense of honour, and not everyone will be as clement as I.

---

148 *Umschreibegebühr* – a fee for changing ownership. The context is unclear, unless it refers to the contemplated purchase of the Eilenhof (see Letter 86). Alternatively, the fee could refer to the house Nobel bought for Sofie in Döbling (see Letter 139), in which case the letter should be dated 1889.

All of this is meant not as a sermon but as instruction, and it is not about the past but your future. And now, my dear Sofferl, I send you heartfelt greetings.

Your devoted Alfred

I again enclose a draft, for you will need the money to be able to pay. I don't know but I have a feeling you are in Vienna again. Please don't deceive me. It would harm you, not me.

91

3 February [1886]

I am now in a condition where nothing in the world would be more welcome to me than repose, and I must have repose or I will soon run out of energy. But first I must deal with a number of affairs that it is necessary to put in order. I must settle my business relations with a few companies, which have been unpleasant for me so far.[149]

Perhaps I have to return to London, then go to Berlin and Turin or even Rome. But it will take less time than one might expect, and it appears that things on the whole are beginning to take a turn for the better. But how much work it is! I could lose my mind over it. Now I need and must have peace soon. It is necessary because I am totally exhausted, and my stomach digesting food is worse off than my brain digesting problems.

Well, then, my dear little toad, it won't be long before I arrive in the dear imperial city. In the meantime I feel sorry for you, my poor thing, for having so much trouble with servants. But why do you mistreat the good ones you have? You treat people like dogs and dogs like people. That isn't fair, for it is only logical that you will be treated like a little dog in turn. So, dear little child, check your Mayer's Conversations Lexicon (have you read it a great deal???!!!) and read the article on human dignity. Read much and think about it even more, and you will acquire a thousandth part of the indulgence you get from your old man, who wishes you well with all his heart.

Alfred

---

149 In 1886/7 Nobel merged his companies into two trusts, Nobel Dynamite Trust Co. (the British and German companies) and Société Central de Dynamite (or "Latin Trust," combining the Swiss, Italian, French, and Spanish companies).

92

22 February [1886]

My dear, little Sofferl

I am going to London again, and although I have only a few minutes, I don't want to depart without sending you my heartfelt greetings. You never know what will happen, given those never-ending rounds of travel. Fortune's ways are unfathomable. Be a reasonable woman. Stop the stupid nonsense[150] that has always harmed you and develop your good side and the better part of your little heart, I beg you devoutly.

Your only true friend,

Alfred

I urgently need to travel to London this time, but I hope that it is the last time I have to rush there. I will stay only two days.

93

[February 1886]

My dear Sofiecherl,

Why do you write so rarely now? As I hear from Vienna, there are good reasons, and I would like to warn you. I mean well. Be completely honest and straightforward with me, or you risk losing the only support in your life.

I couldn't write you much recently. When I come home at night, I am dead-tired most of the time and have to lie down, and so you can imagine that there isn't much writing going on. For some time now I am very, very fatigued. If I could, I would be glad to relax and sleep it off. I wish it with my whole soul.

I suppose you got my letter with the enclosure. Imagine when I wanted to post it, there were so many people at the post office that I couldn't get it done or I would have missed my ride to Sévran. Two days later I found the letter in my pocket. I had completely forgotten that it had not

---

150 Nobel uses *Firlifanz*, Viennese dialect for "nonsense,"

been posted and therefore immediately sent you a telegram to let you know that it was in the mail.

I believe you need to pay a small sum to the bank in March.

How are you, dear Sofferl, and what are you up to??

I would be on my way [to you] but then I would immediately have to return. I prefer to stay a few more days and to arrange it such that things can proceed without me from thereon. For that purpose I summoned Liedbeck[151] here. The devil only knows where he is hiding out. So far no trace of him.

I didn't send back the estimate. I want to order something more reasonable.

Was habitzen vor?[152]

Many heartfelt greetings and 14.[153]

Your old Alfred

94

March [1886]

My dear little Sofferl,

I can't tell you how sorry I am about letting the 1st of March[154] go by without sending you a little telegram. But my sin was not deliberate and can be easily pardoned because I am forgetting all sorts of things now, and it is fortunate that my nose (with all its blackheads) is fixed, or I would forget it every day. You know very well that I am very reluctant to let a birthday or a name day of my old mother pass without sending her congratulations, but I forgot even that a while ago. That is the result of the awful stew I'm in, so that the crowd of thoughts in my head make me quite thoughtless.

You say I don't answer your questions, but I will if I don't forget. It has to do with my forgetfulness, I don't know what's going on in my head.

The devil only knows where that wretched Liedbeck is hiding out. He should have been here long ago to take a load off my work, and

---

151 For Liedbeck see note 3.
152 The phrase is a contraction of *Was habt ihr jetzt vor* (What do you have in mind now?)
153 Code? See Letters 71, 86.
154 Sofie's birthday.

now he is stuck en route for all time and eternity and keeps mum. In this way the days and weeks pass without me being able to get away. I am stuck here because I have responsibilities vis-à-vis the authorities, which have to be taken care of.

All that is very sad because I urgently need peace. My digestion and my ability to sleep are pretty much gone, and the rest of the wheels will come to a stop soon. But it can't take much longer until I come. In the meantime my little toad must behave herself and be reasonable. Sofferl reasonable! That makes me laugh, and that's the nice thing about you – the complete absence of reason. It's gone without a trace.

Sincere congratulations on your birthday, although they are much too late, and fondest greetings,

Alfred

95

March [1886]

My dear, cute Sofferl,

Don't be so discouraged, it will all come right. Every day I feel a greater need for quiet and care and do everything I can think of to make arrangements and free myself for a stretch of time. One can hardly imagine all the things I had to put in order and did put in order, and you in particular, dear little child, will find it completely impossible to imagine. Perhaps if we talk in person you will get a little understanding. The reason for my difficulty in getting away quickly is that everything is in a stew and I am dealing with people I have to treat like schoolboys, that is, I must watch them closely.

It won't be long until everything is in order. And now beautiful spring is in the air with rays of sun and hope, so that my little toad must not despair and instead keep her good humour. Well, there is some benefit in being a little lonely and deserted: one learns to do without caprices and realizes what one owes to other people, especially when they have a gallant mind.

And now it is already 1:30 a.m., and I will try to sleep as well as I can, and wish you too a heartfelt good night.

This morning I sent you a short letter, a few lines only, with a blue piece of paper[155] enclosed.

Alfred

96

[March 1886]

Dear little toad,

You are a thousand times right when you say one mustn't work so hard. I now rarely have time to eat and am no better off as far as sleeping is concerned. I am totally fed up with this business and want to obtain a little freedom for myself, by force if I can't do it any other way. But so far I am stuck here and waiting for Liedbeck's arrival from Stockholm. He is supposed to take over the work in my absence, and deal with my nephew who will come here in an important matter. All that is very unpleasant, but I never needed peace, absolute peace, more than now.

I have no time to write today because I need to go to Sévran, but I add a little piece of paper. Most heartfelt greetings,

Your old Alfred

I am terribly worn out.

97

12 March [1886]

My dear cute Sofferl,

I can vividly imagine how you are feeling and how much you need to talk to someone. You must regard this difficult time as a penalty for sinning so gravely against me, when you had understanding only for your own feelings, not the feelings and moods of other people.

Experiencing problems yourself makes you thoughtful and improves your ability to think. But do not despair, for I feel sorry for my poor,

---

155 I.e., money.

clueless child. Now you have to give up empty fantasy, empty ornaments, empty pretensions, and have to listen to what you owe others
and how one has to behave to earn respect.

I advise you to keep Marie[156] there until I come. She is ill-mannered
but the reason for that is that you are ill-mannered yourself. You can't
demand a higher moral standard from servants than you maintain
yourself, and if you don't display refined behaviour in every way and
at all times, they will take liberties. That is a universal law you can't
overthrow. I will come soon, and until then it would be best to have
someone about you who is not a total stranger.

Many, many heartfelt greetings,

Alfred

98

16 March [1886]

My dear Sofferl,

The business drags on, but it can't be long until I can free myself, and
the best part is that I will then be able to get away for a longer period
of time. A botch-up job[157] is no good, and to run away now with my
work half-done means I would have to return immediately and live in
constant nervous tension. My plan this year is to go to Franzensbad,
preferably in the spring, or to do something similar that would allow
me to put my stomach in order.

Liedbeck is coming, and so I have to interrupt my writing, but send
you warm and heartfelt greetings,

Alfred

After rain comes sunshine, and after all that strenuous work I hope for
a period of quiet and contentedness.

---

156 For Marie see note 49.
157 He uses the Viennese dialect word *pfuschen*.

99

25 March [1886]

Dear Sofferl,

I can't write much because I have been sick now for five days, suffering from a serious bronchial catarrh. Today I am a little better, probably because I consulted a doctor other than Bonato,[158] who always puts on blisters. The other doctor treated me with poultices and quinine, and with a spray of carbolic acid.

Even if everything goes well I will hardly be able to depart before Friday. I needed that in addition to everything else – to become ill! As if my work wasn't taxing enough.

With heartfelt greetings,

Your old Alfred

I am writing this lying in bed.

100

28 March [1886]

Sofferl, Sofferl! I have warned you so often against such unreasonable waste, and don't want to chew the same cud and return to the same subject all the time. Three months have barely gone by and, adding what I enclose in this letter, you have gone through 29,810 francs.

Knowing you, I almost believe that you have saved nothing of that money, yet it would appear that all of Ischl is pensioned off at my expense. That amounts to sucking me dry. And it is so stupid. Really, people should understand that it can't go on in this manner.

My health is not at all as I wish it to be. Bonato thinks I should stay in bed for a few days. But people bombard me with telegrams, letters, contracts, etc. etc.

---

158 Dr Vio-Bonato, whose office was at 79 Rue Lafayette, wrote to Sofie on 6 November 1890 (unnumbered, among her letters to Nobel), answering her inquiry after Nobel's health. According to the doctor, Nobel was then suffering from "une neuralgie à la tête."

Other people are homesick, I am "awaysick," although by God I don't know how far I still have to roam. In Berlin I am urgently wanted, in Turin and Parma even more urgently – I live as in a fever dream. Soon, however, the companies will take all that in hand, and I will enjoy some peace and liberty again – I won't say "enjoy happiness." I lack something that I could never find and likely will never find: a home in a circle of cultured and honourable people who suit me.

Heartfelt greetings and wishes for everything good and dear,

Your Alfred

Don't allow them to fleece you (or rather me) on all sides.

101

31 March [1886]

My dear Sofferl,

My recovery is strangely slow this time, and Bonato thinks I must stay home a few more days. That's just the right thing for someone like me with all that work resting my shoulders. I thought Liedbeck would be able to help me somewhat, but I find him even slower than Fehrenbach.[159] It seems that the North makes people sluggish. He has been in LeHavre for two weeks now and has achieved practically nothing. I would have made more progress in one day. It's not that he doesn't have the will, and he is intelligent, too, but it is hard to produce anything in these innovative affairs. How much easier it is to spend money in this world, especially other people's money, than to achieve something!

Tomorrow I have to go out, whatever Bonato says. I must be present at the official tests, and on Tuesday I hope finally to be able to depart, but not directly to Vienna. First I have to go to Berlin. I have obligations there that I took on already two weeks ago and that had to be delayed on account of my illness. Then I don't know whether I have to go to Italy first or can go to Vienna directly. That, too, has to do with obligations, and everything is urgent. These things are no fun.

---

159 For Liedbeck and Fehrenbach see note 3 and Letter 11.

What a torture it is! And my health is so bad, and my digestion in such a wretched condition. Peace, peace is my whole desire, and peace, peace is the last thing I am granted.

I am tired, dear Sofferl, or I would write about other things, but I must lie down. Even the shortest letter tires me out, and yet I must write so many every day.

Many heartfelt greetings,

Your old Alfred

102

27 April [1886]

Dear little Sofferl,

The first few days here it was very cold and rainy, and the bit of health I had went down the drain. The quarters in Avigliana[160] are so bad that I have to stay here in Turin and must drive back and forth every day. It is very time consuming. Also, Turin isn't Venice, and there is no little toad here.

What has been done so far leaves more to be desired than I thought, so that progress is slower that I hoped. The awful thing about innovative procedures is that one cannot predict when they will be concluded.

Too bad I could not stay in Venice longer. Staying there would have been much more beneficial, providing I wasn't fed to excess.

I can't write many lines to you today. Liedbeck sticks to me, so much so that I can never be on my own. But I send you very many heartfelt greetings, my little, witless, wasteful toad.

Your grouchie-bear

Although the enclosed piece of paper isn't blue but white, you mustn't regard it as an ordinary bit of paper and throw it into the wastepaper basket. It is worth approximately 2,500 francs. Take care of it therefore!

---

160 Avigliana, near Turin, is where the factory established in 1873 was located.

103

London 20 May [1886]

My dear little Sofferl,

Once more I'm riding around the world. Difficulties developed once again with the contract that has been signed but not ratified by all parties, and so it is necessary for me to go to Glasgow to arrange the matter personally. But this must not and will not delay me long. I expect to be back in Paris by the end of the week, where I'll stay as long as necessary to pack up and give the necessary orders about housekeeping, and will then immediately embark on my journey to Vienna. They urgently need my help in Pressburg as well.

I have to admit that I am heartily fed up with work and plan to rest completely. It is no longer work. It is a torment that has affected my health. I am pulled in all directions as if I were made of rubber.

I have to go to the railway station now, my cute little Sofferl. Letters to Glasgow, St. Enoch's Hotel, would not reach me in time. At the very most you can send me a telegraph there, but it would be better to put the Paris address on your letters.

Many heartfelt and fond wishes,

The old Alfred

104

13 July 1886?[161]

Sunday

Dear Sofie,

The lawsuit is keeping me here still, but the day after tomorrow I hope to travel to Stockholm, where I have to look after important matters. My health and my mood are like this summer, i.e., terribly bad. Today I need a coat suitable for autumn, and in the evening a fur coat would

---

161 If the date is correct, the year cannot be 1886, when 13 July fell on a Tuesday. "Sunday" would indicate 13 July 1888. The "lawsuit" mentioned in the first line of the letter would then refer to the so-called Cordite Case. See Letter 169.

serve the purpose best.

Today, Sunday, I sat at my desk all day, but I can no longer work as fast as I did. Old age, worries, and problems of every kind are stronger than the strongest will. And from now on there will be a decline.

In spite of the lack of time, I went shopping for your things. Isn't the old fellow nice? A grey angel, let me tell you, and I shudder at the thought.

Virat's is the most shameless place I have ever seen. It shows what a mad and crazy and good-for-nothing world the financial world is. Imagine, for your little capote hat, which is a mere nothing, I paid 145 francs. Ten steps further down the street, one could buy an equally pretty (if not prettier) hat for 30 francs. Virat's has a lot of black hats now. One should throw those round capotes out of the window, together with the salesgirls and the customers.

Many heartfelt greetings,

The old Alfred

105

[1886]

Dear little Sofferl,

I am still sick and yet keep on working so I can finally get out of here. All those innovations cause so many problems that one cannot predict anything. What torture a poor man must undergo before he can create something!

What I write isn't very coherent, but I am so tired that my thoughts fall asleep as I write.

You never write with whom you socialize and nothing about how you spend that great sum of money, excuse me, I meant to say "waste" all that money. Let me tell you, if you had to work for those blue, inconspicuous pieces of paper yourself, you would discover that one must make economies and can live quite well on less. But that will all come by itself eventually. There are such problems afoot in Russia and England that my income will soon be very small.

And now my time is up, and I conclude with heartfelt greetings.

Your devoted Alfred

106

[1886?]

Dear Sofie,

I am sick, tired, and depressed. Under such circumstances you can easily comprehend that I long to be surrounded by people other than paid domestics. But I can't live with you, my dear child. The past is proof of that. You are too spoiled, or rather your upbringing has been spoiled. You have no understanding of what is owed to one's fellow human beings and often you regard your monstrous caprices as a universal law. That would be somewhat more bearable if you had wit, culture, or skills to balance your whims, but there is no great supply of those, it seems to me. Every time I think of how you behaved at Sévran, I am sort of disgusted. It was understandable, was it not, that I did not want to make the servants there privy to my relationship, and nevertheless you made a face that I cannot wipe from my memory. Only a malicious person would look that way.

Add to this your enormous lack of consideration. You are no longer a child and must surely understand what terrible injustice you commit against me. That can never be remedied. And – have things improved? I doubt it very much. Indeed for certain reasons I got the idea that you had a companion on the journey to and from Linz.[162] May your conscience – if it awakens for once – judge how disgusting such an action against me is!

Trust between us is impossible – you have done plenty to make sure of that. May God protect you from others who might let you taste the kind of injury you did to me.

Did not a certain "gentleman" offer to return for cash the letters you wrote him and did he not threaten to use them to harm you? I wouldn't be surprised if that came to pass or hasn't happened already.

My health is poor. I can't stand it much longer. But where am I go? I don't know, perhaps under the earth!

Sincere greetings,

Alfred

---

162 City in Upper Austria.

107

Berlin, 5 January [1887]

My dear, little Sofferl,

I will return to Paris this evening, but can write only a few words to you because Berger and Hoffer[163] don't leave me in peace. I have just received your sincere lines, addressed to me here, and reply with the same sincerity. I, too, am uncomfortable alone among strangers, but what can one do? All of life is a long battle for achievement and if those who can did not add their contribution, nothing at all would be achieved, and humanity would still live in forests like wild animals. Mind you, in your family they contribute nothing or little towards producing something and a great deal to dispersing it. If everyone acted like that, we would all be in poor shape.

I hurry to Paris, because I urgently want to see what Fehrenbach has prepared before he sends it off. He is so distracted and so thoughtless that I fear something is wrong with him.

Here I have smoothed things over as far as the short time permitted it.

Now take care, my dear little toad, and likewise put everything in order as best you can and write to me in Paris a little report, but step carefully, don't rush anything, and most importantly, don't get angry. Show some consideration for people as well, because you yourself need their consideration in turn since your poor judgment is not well rooted and is often twisted awry.

From my heart, and with the fondest greetings,

Your Alfred

108

12 January [1887]

My dear little Sofferl,

You cannot imagine how busy I am now that I am back. People pull me in all directions so that I don't know whether I'm coming or going. And

---

163 I cannot identify Berger. For Hoffer see Letter 11.

now I have to run around because of your boxes and suitcases.[164] I will do it as soon as possible, but I absolutely can't do it today – I don't have a single free moment.

Enclosed you will find some rather essential bits of paper, which are hard to come by. In addition, many, many heartfelt greetings.

Your Alfred

109

[1887]

My dear little Sofferl,

I'm so sorry for you. Really, I don't think there is any evil in you. In my opinion you have sinned against me so much and so gravely through your stupid lack of reason. There was no end to what you expected me to suffer and put up with. And now you feel the consequences. If you weren't the little, weak, senseless, undecided creature you are I would have left you in the lurch and would have disappeared after living with your stupid and capricious behaviour for one week. But as it is, your weakness served you as a good defence.

You are anxious, my dear, weak child, because you are lonely, but you don't understand that I can't stand being with you exactly because I have nothing to do. Either I have to have true company, that is, someone who offers me intellectual food, or I must occupy my intellect. I always missed that when I was with you, and that is why you believe I am an inconstant person who can't stand any place or any person's company. That is completely wrong, and no person is less demanding and less changeable than I. But I don't want to become a vegetable, nor associate with someone who cannot understand that one has certain social duties to fulfil.[165] I don't have to observe all social norms, but I

---

164 The context is unclear. It may be connected with Nobel's plan to sell his villa in Ischl. He asked Sofie to vacate it by the end of July. Her father protested in a letter of 28 July 1887: "Think of the scandal this would cause ... after an intimate relationship of ten years" (Appendix, p. 278). As a result of Hess's letter, Nobel allowed Sofie to stay in the villa until September.

165 The remark about "social duties" is best understood in the context of a visit Nobel received at this time by Bertha von Suttner and her husband Arthur. Nobel took the Viennese pacifist (who had briefly been his secretary) to the house of Juliette Adams, whose famous salon attracted many intellectuals and artists.

must adapt to them to the point that I don't offend others and make myself ridiculous.

All that is Greek to you. You can't understand it. And that is lamentable, for how can there be an understanding between us in life, when I can't clearly communicate to you the first and foremost point of honour?

Now I have to go to the train station. Greetings from my heart, my poor, little, abandoned Sofferl.

Your fondly devoted

Alfred

110

[1887]

My dear little Sofferl,

Don't be in such despair. Raise up your little heart and take fresh courage. It appears that you see clearly at last how stupidly and improperly you have acted for so long against a person who is made of pure benevolence and philanthropy, who thinks so little of himself and so much of others.

I too am worried, my poor little child, because I have become so terribly lonely, and my present work is of a kind that offers me no peace and no distraction. You would not believe how empty I feel and how much my health suffers under that awful work. Sometimes I come back from Sévran and must go to bed at once, and day before yesterday I almost fainted in Sévran. Sometimes I feel like leaving everything behind and settling down in some small village to enjoy peace in my old days at last. I love work, but not the kind of work that destroys body and soul.

Indeed you are lucky, my little toad, that you have no sense of honour because one has to suffer a great deal for it. Yet I don't want to lose it, indeed it is the only thing to which I am attached with all my soul.

And now, my little child, don't worry. Don't torment yourself with bad conscience about all the sins you have committed against me. Although I cannot forget them, I will forgive them. Live and sleep in peace, dream of better and happier days, and accept my heartfelt greetings and embraces.

Your old protector Alfred

Actually, I don't understand what kind of certificate you want for the remaining things.[166] In my opinion one should either sell them, because they are mostly things that are not worth sending on, or send them to Ischl in the summer, where they don't make a fuss at customs (if you recall). What do you think of my proposal?

I enclose something for your balance.

110a

[April/May 1887][167]

Dear Sofie,

To avoid unpleasant discussions in Vienna I did not express my full sentiments about that incident on the Ring.[168]

When you stayed at a second-rate hotel[169] instead of the Hotel Meisel [sic], it was clear to me that you had resumed your affair with the gentleman from Pest. That became even clearer when that gentleman had the nerve to address a letter to me, and it became crystal clear on the Ring.

The story about the old woman was a brave invention and shows that you have made significant progress in lying.

But I am uncomfortable with all of that. I prefer to live a lonely life to being constantly lied to and deceived and ridiculed. In any case you will discover – and it will be a bitter discovery – that most of the time you deceive yourself with such actions.

I can excuse much if I am told the truth, but if I am confronted only with lies, I am disgusted.

---

166 See Letter 108.

167 The letter is dated 1888 in Arkiv ÖI-5. I reassigned it to the year 1887 because the contents match those of a letter from Heinrich Hess of May 1887 (see Appendix, p. 277). In that letter Hess refers to an "unpleasant incident" and then makes excuses for Sofie's continued contact with H ("the gentleman from Pest" mentioned by Nobel here; see next note).

168 The Ring[strasse] is the grand avenue ringing the inner city. The incident must be connected with Dr Hebentanz of Budapest, a physician elsewhere referred to as "Dr H" (see Letters 80, 153). According to Sofie's father, Nobel had at one time expressed a willingness to arrange a marriage between the physician and Sofie, but changed his mind when the man was revealed as a gigolo. See Appendix, pp. 277–8.

169 Presumably a reference to the Hotel Sacher. See Nobel's deprecating remarks in Letter 111.

There is an abyss now between us, not of one but of a thousand fathoms. You ascribe the change in me to others, but you are completely wrong. I associate with no one and am so lonely now that I never even go out to dine. You think Olga is of interest to me.[170] She does not interest me personally or as a woman but simply as an experiment. Can one successfully correct such a thoroughly corrupted child and turn her towards a better life, at least for a while? I am not interested in spiteful people. I don't know why – I dislike them intuitively. You would have seen all that long ago if you could only understand that one may help a person without selfishness and ulterior motives. Among the Israelites, this occurred only to one man – Christ – and because it is so rare, he was given a diploma of divinity. If you want to live in Vienna, you must rent a place yourself in your or your father's name. It would be nice, wouldn't it, if the gentleman from [Buda]pest would one day live in accommodations rented in my name.

What do you want to do in Paris – I can't explain it. Perhaps you consider it a good opportunity to spend more money, and there would be no limits to it. Certainly you won't find me here, because I now live in Sévran and will very soon go on a long, long journey.

Best,

Your Alfred

111

October [1887]

Dear Sofie,

It must be written in the book of fate that you should spoil my good will. That crafty action of your former admirer,[171] who sent me an anonymous telegram, would have helped you a great deal in my eyes if you had not spoiled it with your own lies. Your letter from the day before yesterday contains several lies, which are exceptionally stupid since

---

170 See Letters SH 34 and SH 37 for Sofie's jealous accusations. Heinrich Hess, answering Nobel's accusations on her behalf, alleges that Olga slandered Sofie. See Appendix, p. 279.

171 I.e., Hebentanz. See Letter 110a.

you must know that I am well informed. You think it was that fellow from [Buda]pest who informed me, but that is an error – I heard everything from another man, who does not lie.

Now you have gone once again and without my permission to the Sacher,[172] which is a hotel for single men. I assume therefore that you will pay there without my contribution. You are out of bounds and rely too much on my patience. It is stupid of you. Where in your life will you find support if you lose mine?

I have been ill now for nine days and must stay in my room. I have no one around except a paid servant, no one who asks after me, and all that is the result of my acquaintances snubbing me on account of the nasty gossip caused by your behaviour in Vienna, where you use my name to which you have no right. And after all this you have no qualms about asking me for money – that is fantastic!

Since I did not permit you to stay at the Sacher, you can pay for it yourself.

I believe I am much sicker than Bonato[173] believes. The problem is persistent and will not go away. If a man is fifty-four years old and all alone in the world, and a paid servant is the one who shows him the greatest benevolence, sad thoughts will occupy his mind, sadder than most people believe. I can read in the eyes of my servant how sorry he is for me, but of course I can't allow it to show.

Many heartfelt greetings,

Alfred

---

172 Hotel Sacher at 4 Philharmonikerstrasse behind the Vienna State Opera was established in 1876 as a rooming house and, from 1880 on, run by Anna Sacher, notorious for smoking cigars and surrounding herself with French bulldogs. Nobel clearly disapproved of the hotel, although it was frequented by high society. Nobel had sent Sofie a telegram (quoted by her father, Appendix, p. 280) to stay at her parents', or perhaps at the Hotel Meissl until suitable accommodation was found. Sofie's father defended her against Nobel's indignant remarks, saying that she was eager to leave the hotel and find a permanent home (ibid.).
173 For Bonato see Letter 99. Sohlman, *Legacy*, p. 46, refers to him as "Dr Bouté."

112

31 October [1887]

Dear Sofiechen,

Of course I can't go on living the way I do for any length of time and will have to arrange my life differently. It is the ninth day today that I am sick, and no human soul cares about me with the exception of a clumsy man who is paid for his service. That is hardly a return or reward for all my efforts and all the benevolence I show and have shown to my fellow human beings.

I thank you very much for your kind offer to come here, but I know what it means. It means you will expect me to run around in spite of my sick lungs and look for accommodation for you, and to listen to your constant complaints that it is not good enough, and to you pouring out insults and unjust accusations against my relatives, and I would have to listen to slander, endless ugly scenes, boundless pretensions, etc. All that is hard to take even for a healthy man, not to speak of a sick man who furthermore suffers from melancholy.

I admit that there is something nice and loving in your nature, but when you are with me, that side is hardly in evidence any more. For years now you do nothing but deceive me and lie to me and in a way that is not only hurtful but also harms my reputation in the most awful manner. And in addition you come up with silly excuses, saying that a name has no reputation. That expression in itself shows the strange company you keep and what nice things the people teach you.

In spite of all of this I cannot forget that you were once fond of me. But for years now not a shadow of your fondness is left, and even birthday wishes usually conveyed by polite women fell by the wayside this year. But in former times, before you entered the rotten atmosphere of certain admirers of yours, you had something truly womanly about you, and remembering that, I overlook and forgive some things. I have another reason as well for pardoning you: I certainly don't consider you fortunate. In my opinion you have left the right path, living the way you do now, and waste your best years on a miserable life in hotels that would disgust even a man, not to speak a woman. You constantly get involved with admirers you pick up God knows where and how, who give you little joy or contentment. They speak ill of you afterwards and occasionally lie to you and cannot but leave you in return with a disappointed heart.

But that's what you were asking for, and instead of creating a home for yourself, where your good traits could attract friends in time, you widen the chasm between us through your thoughtlessness, not to call it anything worse.

Your benevolent Alfred, who always means well with his whole heart

---

113

5 November [1887]

Dear Sofie,

The way you waste money is not only a sin but almost a crime when one considers the many poor fellow creatures who have a large number of children and must starve, while you absolutely throw away money. And – do you get anything out of it? You have none of the good and pleasant things life can offer, and yet you waste so much money that twenty families could live on it and be happy.

My health is very poor. My bronchial catarrh is a little better, but I suffer from neuralgia as never before. I have become quite weak and am terribly run down.

Sincere greetings and a thousand reproaches, which will have no effect, alas.

Your old Alfred

114

13 November [1887]

Dear Sofie,

You have no talents and little brainpower, although you don't lack feeling and have a loving nature. If you had not sullied my name in this manner, we might have gotten along alright. But your lack of consideration worked against it, and that's how it is: The sins we commit against others come home to roost.

It seems your life is not happy since you gallivant around in Austria and do everything in your power to ruin the reputation of an honest man. It's downhill for you, step by step, and soon, very soon, when I am in my grave, who will say a good word for you or do anything good for you? You alienate the whole world, and yet, fundamentally, you have a better heart than others. But spite without judgment always makes for a bad situation, and you will create a true desert around yourself. Poor, poor Sofie! You could have done so well, and now what is your life like? Sad, isolated, nomadic. And no one feels sorry for you, except I alone. The others see only your great sin and your mistake. They are jealous of your fortune which is a misfortune, find everything in you wicked and wrong, and yet they don't show you a cross face, but only because they think you could be of some use to them perhaps.

And that is how your life goes on without support, without true love and affection, with rouged cheeks, stupid ornaments, and hollowness in your heart and soul. As you can see, I am still grieving for you.

Your Alfred, who is still deeply and sincerely devoted to you.

115

[1887]

My hair has turned quite grey, but by far not as grey as my mood. It is an odd thing when a man at my age, with the grave the nearest neighbour, is so completely alone in the world that no one cares for him except an old servant, who helps undress him. What's more, I do hard and strenuous work fifteen hours a day and have all sorts of worries and problems as well, and now my health is completely shattered.

What I find most depressing is the fact that I have absolutely no one for whom I would live or die except you, a woman who considers nothing more urgent and more desirable than to damage my honour daily and hourly and to hurt me – I won't count her. I adopted you, so to say, as an orphan and out of pure and noble considerations. I got a nice reward for it and am still getting it.

No hour passes without my being afraid of gossip. It has practically driven me to solitude, and yet I am not rid of my fear.

It would be very, very necessary for me to stay here, but I am so broken up that I can no longer stand it. I have to go where my eyes can enjoy some green, and where I can stay at a hotel which, however bad, is a thousand times preferable to my present home. I feel nowhere so deserted and alone as here. When I stay at a hotel, the life there always seems like a passing evil, whereas here I feel that my life has settled into permanent bitterness.

If I could only have the company of a person who understands me and was not determined to damage my honour. I might find such a person, especially since I don't care about age or sex, but one must have time to search, and I have less time than ever. In fact I can hardly fathom how I managed to put this letter together.

I just got your telegram, from which I see that you are ailing. I am very sorry. In any case I would be inclined to feel sorry for you, if I did not observe on every occasion that you hurt me and reward all my kindness with ingratitude. We just can't understand each other. Just as a blind person doesn't understand anything about colours, you understand absolutely nothing about consideration and feelings of honour and delicacy.

Sincere greetings,

Alfred

116

1 December [1887]

Dear Sofie,

I am heartily sorry to hear that you are really ailing and hope that it is nothing serious. The doctors always paint everything blacker than it is. I suppose it is only your usual ailment getting worse. But you are so awfully careless, my dear child, and won't accept good advice. You don't have the constitution to rush around. As soon as you are healthy again, let this be a useful lesson and live reasonably and on a diet.

I too am very sick, not exactly confined to bed, but worse because my mind and my spirit are sick, and my digestion has never been worse. The old machine won't run much longer, and because I know it well and sense it, I work like a horse to advance my numerous tasks. But it doesn't go as fast as I would like it to go because I suffer from headaches to the point of torture, and so my thoughts don't run well either.

I enclose money and wish you from my whole heart a quick and complete recovery and the intelligence to be careful once the difficulties are over.

Heartfelt greetings,

Your old Alfred

117

24 December [1887]

Dear Sofie,

I don't think I need to explain that it is unpleasant to spend Christmas Eve in a hotel, surrounded by strangers. Now more than ever I would need loving company, because I have suffered greatly in body and mind recently, and you cannot imagine how exhausted and tired I am. I would have very much liked to go to Vienna for Christmas if two important reasons had not kept me back: first, that I could not have stayed at Meissl's[174] and did not want to upset you by moving to another hotel. Secondly, I could not have avoided spending Christmas Eve at your relatives, which would have been too upsetting and strenuous in my present condition. In addition, I can no longer travel long distances without taking a break. I feel very, very much aged and believe that I am heading for problems in the not too distant future.

I hope, dear Sofie, you have no worries and that your health is completely restored. It was only a few clouds passing through your sky. Perhaps it was good for you. Your wilfulness has finally brought it about that no one can stand being with you, and restraint was urgently needed and in your interest. Did the hard times – well, not so hard, really – bring about an improvement in you? If so, you can thank God for the ordeal, for your life would soon have become a true hell through your own fault. And yet you always had before your eyes the example of a man who is gentle and very forgiving in his judgment of others. In spite of that you turned into a regular Fury, who hardly knew any feeling other than hatred. Oh my dear Sofie, put away the old ugly quarrel in the new year and retain in your young soul only the loving nature, the gentle sentiment, and the cheerful mood that you were born with, and your life will go as well as I wish with all my heart.

Your fond Alfred wishing you well.

Even if the snow, which is now coming down in spates, does not let off, I will arrive there around Wednesday, or in Brünn or Graz.

---

174 See note 172. Apparently Sofie moved from the Hotel Sacher to the Hotel Meissl, as requested by Nobel.

118

18 January [1888]

Dear Sofie, my great devourer of banknotes,

I enclose new fodder with the express and emphatic statement that this isn't hay, and these inconspicuous pieces of paper don't grow like grass but are very hard to come by. Perhaps there will be a time when the loving God will make this kind of fodder rain down from heaven instead of the manna he provided formerly, but I doubt it will happen in our time.

The night journey on the Orient Express again completely robbed me of sleep, and since I parted from you in Vienna, I haven't slept a wink. Perhaps this is partly the fault of the unpleasant situation at home.[175] I gave him only half an hour to leave the house, and the explanations he offered didn't help him. He had to own up. I feel sorry for the old woman, for where will she be able to find good employment now?

I have so much work that I can't write more today and conclude with heartfelt greetings,

Your devoted Alfred

The nice liver looked still fresh but had a strong and pronounced odor and arrived in Paris untouched and must have pleased the conductor as much as it pleased Bella later. I am sure he will pray for the donor. Perhaps the pope will canonize us.

119

18 January [1888]

Dear Sofie,

To speak candidly, I must confess that I liked you in Brünn[176] a hundred times better than in Vienna. Is it the big city air or other influences? In Brünn you were quite loving for two complete days, and I already

---

175 Context unknown.
176 Now Brno in the Czech Republic.

began to believe that there are women with whom one can keep company without being annoyed. But the unbelievable did not come true, and you changed in Vienna, not only in your nature but even in your appearance. Your features became harder and less beautiful. I am not an analyst of human nature and don't seek to fathom the reasons. I merely state the facts. And that's all. I wrote to you by registered mail today and add here only heartfelt greetings.

Your devoted Alfred

120

[1888]

Dear Sofferl,

Today I received a letter from Robert, in which he asked me urgently to go to Cannes[177] in the near future. It does not suit me since my health is deteriorating day by day, and I can't stand the windy spring weather. By the way, I prefer to sit in a train compartment than in my house, which pleases me less than ever. Things were bearable formerly, when the housekeeper[178] kept good order and knew how to arrange things and wasn't a complete mercenary soul at any rate. Now I feel eerie, and as soon as I have some free time, I will make other arrangements.

Today I am quite ill and in a depressed mood. I send you heartfelt greetings,

Your fondly devoted
Alfred

I'm not looking for the ideal woman, but she must have a little tact and understanding for what she owes to other people.

---

177 Nobel's brother Ludvig was staying at Cannes at the time. He was gravely ill. See Letters 122 and 123.
178 He had dismissed her. See Letter 118.

121

[1888]

Dear Sofie,

Here I am in the twilightof my life, without friends and isolated, so that I would heartily welcome the company of a sympathetic person and long for it with my whole heart. But you cannot fill that role, for several reasons. Most of all you lack true sensitivity and an understanding of what you owe to a man who takes care of you and asks for nothing in return except the consideration every decent person owes to fellow human beings. You cannot understand any of that, and now it is too late to think of improvement. What you lack in intelligence and education you were able to make up, to a certain extent, through your sentiments, your openness, your friendliness. Although you knew very well that I lost huge sums over the last years and have no great income, you did not stop your mad wasting of money and seem to think that you can go on abusing my good nature in that manner.

At least learn from your past life that lying and cheating often harm the perpetrator most, and if you can find a person who still trusts you, you may live – perhaps not in complete happiness, but at any rate in modest happiness.

For me the cup is running over and the knowledge that I acted generously and nobly and am rewarded for it in this manner redoubles my bitterness.

As for your illness, it seems quite widespread in Vienna and can always be traced to the same source. Gross,[179] as you say, seems to have suffered similar symptoms several times, but I hope you will not suffer similar consequences, because his health, I understand, will never be restored. Fortunately you seem to be well again, and I hope this will continue, if you don't commit another folly.

I am not angry with you – I've never been one to bear a grudge – and I wish you happiness with all my heart. But you seem to wish that I should trust you as well and completely forget that all those years of cheating and your unspeakable lack of consideration have made that impossible. You once said that I am a "bit brutal." Tell me, which of

---

179 Perhaps Ferdinand Gross (1840–1904), writer and president of the Vienna Press
     Club, Concordia.

us acted brutally? I leave the answer to your conscience and send you sincere greetings.

Your old, more than upset and more than hurt

Alfred

Everyone has betrayed me except Thomas,[180] so that I must be beholden to him and not trust anyone more than natural reason tells me they merit.

Be completely honest and write without holding anything back how you want to arrange your future so that you may have as bearable a life as possible.

122

9 April [1888]

Dear Sofferl,

I have just come from Cannes[181] and am so fatigued that I can barely hold the pen.

From the plan that I herewith return to you it appears that there is only one servant's room and a very small room, or could one use the attic for additional rooms?

My brother is ill with a grave heart disease and may not recover. The doctors find it rather puzzling that he is still alive. Is appears that I too am not completely free of this problem. My bones won't get old. It's not a grave matter though – no one will grieve over my death, least of all those whom I have helped.

More tomorrow. Today I am too tired.

All the best from my heart,

Your Alfred

---

180 Context unclear. Is the reference to Thomas Reid, who succeeded Alexander
      Cuthbert as manager in 1883? On Cuthbert see Letter 73.
181 He visited his brother Ludvig, who died there on 12 April.

123

3Paris 13 April [1888]

Dear Sofferl,

My poor brother [Ludvig] passed away yesterday after a long and grave illness. He died a gentle and, it appears, painless death. Since the funeral will take place in Petersburg, I did not return to Cannes. Instead the family will stop here en route [to Petersburg] and rest for a few days, staying with me.

I just received your long letter about the plan, but I can't reply today. I am too busy and very fatigued.

Are all toads as silly and thoughtless as you? You are no longer a child and should be advanced enough in mathematics to count to 28 and pack ahead of time.[182] That would be better than to lament afterwards. You should teach Bella the great skill of mathematics because there is no lack of intellect in that black-and-golden old, grouchy, and crafty spinster. She is ten times more virgin than Mary, and if there is a Holy dog-spirit, he will be tempted! It would be desirable because the canine tribe is becoming more and more immoral and needs redeeming.

Best,

Your Alfred

124

18 April [1888]

Dear Sofferl,

I am very busy so that I couldn't find time to write to you earlier. For some time now I am so tired in the evening, especially when I return from Sévran, that I go to bed around 9 or 9:30. Thus even more work piles up and that often makes me so nervous that I can't work for that reason.

My nephew and niece will come only in June because the young man must pass his exams first, and so I am still without a soul, and really it's better so, because otherwise my time would be even shorter.

---

182 Point of reference unclear.

Really, it would make the most sense to depart immediately and enjoy the beautiful spring days, but then what will become of all the work I have started?

I enclose something truly Hungarian, without paprika, yet useful and necessary for your household expenses.

I address this letter to your Papa because it seems that such a Hungarian message is safer that way.

Loving wishes, from my heart,

The old Alfred

125

19 April [1888]

Dear Sofiecherl,

Yesterday I sent you 3 Hungarian papers, worth about 6,600 francs. I enclose four blue papers now. You must get rid of all your debts.

I still don't have peace and no time to write. My health is not good. That is probably the fault of all that work and also of the bad weather. And old age!!!

All the best from my heart,

Alfred

---

126 is now Letter 110a

127

Vienna July [1888]

Dear Sofferl,

I came here because I promised to come and to inspect the place you told me about in Währing,[183] but the journey took such a toll on me that

---

183 Then a suburb, now part of Vienna. Nobel has agreed to pay for a villa to be acquired by Sofie. See also Letters 128, 129 for efforts to find a suitable property close to Vienna.

I could only write these lines with the greatest effort.

The place is not bad, but it has a number of drawbacks. The basement is exceptionally damp and it will hardly be possible to dry it out even at great expense. The staircase is well lit but not particularly beautiful – not a bit of marble, as you nattered. The garden could be pretty but isn't at present because there are no trees and the bushes are nothing special. The rooms on the main floor are pretty and comfortable, but there is a lack of rooms for servants because no one could live in that miserable wet hole below. In any case the woman is asking much too much for such a small place. She wants to retrieve everything her late husband has wasted, and probably also everything his lordship ate.

It was as I thought: When you and your brother[184] are asked to give a report, nothing useful can come of it.

In my opinion, the place you inspected in the Heugasse is incomparably more comfortable and more pleasant in every respect. But you will dawdle and go back and forth until I'm fed up with continuing to support you, or death will deprive you of my support. Perhaps that time is not far away. Shortly after your departure something happened that may be quite uninteresting to others but clearly showed to me how unfortunate it is to have no one around whose loving hand will once close my eyes and whisper a true word of comfort. I must find such a person, and if there is no alternative I will move to Stockholm and stay with my mother. There is at least one person who does not want to fleece me and who brings no caprices or moodiness to the sickroom.

And now the incident I mentioned: Around 2 a.m. I suddenly felt so ill that I did not have enough strength to ring the bell or unlock the door. Thus I had to spend several hours, all by myself, without knowing whether they were my last. Heart spasm apparently, which I have suffered earlier, but not in the chemical laboratory. After that experience, I am discouraged and my heart is edged like this paper.[185]

Now I would like to give you good advice, earnestly and from my heart. Arrange a [permanent] home for yourself. This awful galloping around is against nature. Nor can I understand why you don't finally realize that you mustn't use my name without my permission. That prevents me from being as benevolent towards you as I would otherwise be, and there will come a time when you will feel bitterly the

---

184 Sofie's brother Luis Hess. For the Hess family see Letter 132 note.
185 I.e., edged in black. Nobel was in mourning for his brother Ludvig. See Letter 123.

injustice you inflicted on me by your action. Either I am married, in which case I have no reason to conceal it, or I am not, and then it is shameful when a woman appears under my name. I am no impostor and don't see why I should appear to be one through the actions of others and my own benevolence. But that does not concern those who think only of themselves, but even they will one day find it painful to think of the disgusting injustice they committed against me.

I won't and can't go to Ischl. I don't need to tell you why.

Mr Gschwandtner[186] wrote to me asking me for money. I did not answer him since you were already en route. I suppose you will put the matter in order.

I enclose a cheque.

Many and sincere greetings from your Alfred who is in a sad mood.

128

Vienna [July/August 1888]

My dear Sofiecherl,

I have decided to stay here until this evening in spite of the oppressive heat. The reason is this: I am trying to find for you a nice home in Vienna or rather very close to Vienna. Although I am a complete stranger here, I found in one day better and much cheaper places than your Papa, your brother, and you could find in a year. That doesn't say as much for me as it says about the skill of others. I have excellent prospects in Dornbach. It is rather distant from Vienna, but less than ten minutes farther than the cottage in Währing, and what clean air there is and what beautiful landscaping! It takes 20 minutes to get there from Vienna, that is, from the Ring, and that distance is no greater than, for example, from the Theatre de la Porte St Martin to my place. I also saw another villa there, which is not beautiful from the outside, but is really cozy inside, and with such a splendid garden that I was amazed – and I am not easily amazed. And all that was offered at prices quite different from the villas your relatives selected. In short, one can find in Vienna what would be very difficult to find in Paris and only at a horrendous

---

186 For Gschwandtner see Letter 68.

price. You can see, dear Sofferl, now that I have taken the matter in hand myself, I am even more perplexed that you hung about in hotels in Vienna for so long, when it is so easy to establish a pleasant residence that is also a comfortable home.

Take good care of yourself, put the brakes on your bad mood when it overcomes you, be a reasonable little woman, or if you can't be that, be at least consistent and wear short dresses so that your attire is in tune with your childishness.

Yesterday I also saw a pompously furnished but not very comfortable house, which a very well known young prince established for his mistress, they say. I didn't like that story, however, and perhaps the mistress didn't either for she sold the house a long time ago.

And now, dear Sofferl, I continue on my tour of inspection and say a heartfelt adieu and kiss the paws of Bella.

Your Alfred.

Tuesday morning, 7 a.m.

129

[Berlin, August 1888]

My dear Sofferl,

I have just arrived here, very tired. And now I find here telegrams from Robert and from my nephew, in which they practically beseech me to go to Petersburg and tell me that my presence there is absolutely indispensable.[187] I myself have the feeling that I could perhaps prevent an infinite stupidity there. But I am so fatigued that I am very unwilling to embark on such a long journey, which is torture.

In Vienna I saw a number of villas and houses. Among them, the one in Dornbach is especially beautiful. It is not completely finished but the plaster is dry already, and it will turn out exceptionally beautiful. Both garden and house are very pretty. The journey from Vienna took me

---

187 I.e., in the wake of Ludvig Nobel's death. His son Emanuel was to succeed him in directing Branobel, the oil company in which the brothers held shares.

20 to 25 minutes (to the Hotel Imperial) and it is less than ten minutes to the Dornbach forest, which is splendid.

The villa cost over 75,000 gulden and can now be bought for 30,000, but one would have to decide quickly because I think it will be sold before long. One might have to spend a few thousand in addition, but then one would have something really nice.

If you want fresh air, there is nothing closer to Vienna than Dornbach and connections are very convenient.

Of course the villa, which is unfurnished, does not look as good as rooms that are lived in, but I am sure that the house would turn out much more beautiful than the cottage you liked so much. The address of the villa is 14 Pichlergasse, Dornbach. If you like the idea, you could perhaps go on a day trip there from Vienna and inspect the villa. But before completing any transaction, it would be advisable to have an architect inspect the villa and give you an estimate of the cost to complete the remaining details. Don't make a fuss about the way the stairs are constructed because that is the fashion now, and it offers a certain convenience, too.

Well, then, my dear Sofferl, make the final decision. I am making an effort to find a pleasant home for you and instead of appreciating the benevolence I am showing toward you, you only make new troubles for me without any reason and create ill feelings, unfortunately not without reason.

The house your Papa pointed out to me in Döbling (Hirschgasse 61)[188] is also nice, although much less so than the one in Dornbach, and the neighbourhood is not pleasant at all: a madhouse, several hospitals, a cemetery, etc. Otherwise it would be a comfortable home. I liked it very much. Your Papa thinks it could be bought for 40,000 florins – yes, get something like this or the house in the Heugasse or any other street, but make it fast and don't waste time again.

I end with many heartfelt greetings,

Your old protector Alfred

---

188 The street number is incorrect, as Nobel himself probably came to realize, since he asks Sofie for the number in Letter 141. I have not been able to find the correct address.

130

[Berlin, August 1888]

My dear little toad,

I seem to have succeeded, by telegram and letter, to bring Robert around to reason, so that he has given up for the time being the idea of dismissing the whole directors' board and voting in his own people. But my return to Paris is urgently needed, and trips to spas are needed as well since my health is not good at all. A single night journey makes me tired for weeks and incapacitates me for any work.

So nothing came of Dornbach. I tortured myself to find something nice for you, but I've decided to give that up for the future, because nothing is good enough for you, like a pampered little princess, and you can never make up your mind.

Now I have to admit openly that I miss the quiet pleasant life in Ischl and really look back with longing, but so much happened all at once this year that I quite shy away from it and don't know how I should manage all the miserable travelling about. If I travel to Stockholm now, I would have to give up any water-cure this year and then the winter will be awful for me.

You haven't written how you are and how you spend your time. If your life is really more pleasant when I am with you, I will do my utmost to get there soon. But in that case I have to arrange to go to Stockholm before the late fall. My mother has reached the age where one mustn't put off any plans to give her joy. And who knows, I may give her much sorrow soon. Lately I often have such a strange feeling around the heart that I am convinced that I have a cardiac defect, in spite of the assurances the doctors give me. But it doesn't matter much. Few people will grieve the death of the old man.

Accept your old grouchy-bear's fond love, who has been forced to growl, for otherwise I am really a lamb,

Alfred

I am about to depart for Paris but will rest in Cologne or in Brussels if I get too tired. I won't travel through the night anymore. That is pure [hell].

131

3 August [1888]

My dear, cute Sofferl,

Your short letters that arrived here gave me joy. As far as the style is concerned, they don't compare to our greatest poets, but they very nicely combine sentiment and cheer. How is that for praising your letter? In the end you will be so proud that you will send a sample of your French epistles to the exhibition.

I don't know whether it is the fatigue of travelling that causes the frequent heart cramps I experience and which make me awfully melancholy. Travelling alone has become very disagreeable to me therefore, so that I tried with every means to settle the business in Petersburg without having to be present there. I was successful to a certain extent at last, and even Robert telegraphed that he would no longer try to persuade me to go there for such a short time. I was glad because the overnight journey to Berlin gave me more trouble than you would believe. On the return trip I had to make a rest stop in Cologne as well as in Brussels.

Brussels is a nice little town that I come to like more and more. The surroundings are so beautiful. But not as beautiful as Ischl, where I was pleased with the good treatment this time rather than with the mountains. You were really very good most of the time and were less capricious during those six days than at other times in one hour. In the end you even displayed some understanding for what one owes to others, and you will see how well that reflects on yourself.

Here construction has begun. My people seem to be quite reliable and everything appears to shape up well. I have more difficulties with Sévran and Barbe,[189] but I want and will go to a spa, and I also long for Ischl. But I can't avoid taking a trip to Stockholm. It would hurt my mother very much, and consideration, fond consideration is my life's task.

Give my regards to your dear brother. Fond greetings to you,

Your old, sincerely devoted Alfred

I am so nervous that I can't write. Excuse my awful handwriting.

---

189 For Barbe see Letter 18 and note.

132

14 August [1888]

My dear Sofferl,

The idle life in Ischl is very calming and very pleasant but it's hardly recommendable as a place for a water cure. Until now only Aix was beneficial for me, and I hope I will benefit from it this time as well. If nothing important intervenes, I will go from there to Ischl and undertake the journey to Stockholm later. Unfortunately the reports I get from there are not positive, but I feel too exhausted to start immediately on the strenuous journey to Stockholm.

What's "doing" in Ischl these days?[190] You say people find you nice and pleasant. What people? Wouldn't it be good to add a little list here, or you'll be subjected to a @#$%[191] cross-examination next! Even Bella's pleading won't help you then.

I had a letter from Gschwandtner, who enclosed a statement concerning the lot. Dr. Melkus[192] suggests I buy a part of the abutting meadow, but I don't think that old barrack is worth such an investment. In any case, the matter can wait until my arrival.

I would like to write about a few more thing to you, but I am about to go to Sévran where I have to put some things in order before my departure for Aix. Yesterday I was in Le Havre to get some tests started and hope that everything works out well without any need for me to interrupt my water-cure in Aix.

Many heartfelt greetings to my little toad, which will turn into a pearl if only reason surfaces and moodiness sinks to the bottom. Blessed be such an almost impossible and in any case unbelievable event. If Malie[193] can manage that, she is a true sorceress. Again, fond greetings without growling.

---

190 *Was "habitzen" jetzt?* Nobel imitates Sofie's Viennese dialect.
191 Nobel uses a quirky Viennese curse: *kreuzdonnerwettrig*, literally "crucifying thunderstorm!"
192 Not identified.
193 Sofie's sister. Heinrich Hess and his wife Amalie (née Störk) had five children: Sofie (b. 1851), Max (died as an infant), Bertha (b. 1853), Caroline (died as an infant), and Amalie ("Malie," b. 1856), who married Albert Brunner. After his wife's death, Heinrich married Julie Stern, with whom he had four surviving children, Ludwig ("Luis," b. 1858), Hermine (b. 1861), Käthe, and Anna (born after 1865).

Alfred

Write to me at Aix-les-Bains, but without mistakes, please.

133

Grand Hotel Stockholm, 18 September [1888]

Dear Sofferl,

I have been sick since my arrival here. Sometimes I even have to stay in bed. My energies seem to be exhausted, and I won't be able to travel soon. I feel especially ill at night, when I am plagued by fever and heart problems. I don't assign much value to what the physicians say. One says "black," the other "white." My own feeling tells me that I am experiencing a steep decline. I haven't received the letter from Vienna, in which you were to tell me about the little house.

Just now Satan[194] is coming to visit me again, likely to sponge off me, and I have to end my letter with heartfelt greetings,

Your old Alfred

134

[1888]

My dear, cute Sofferl,

I can't write much today because since my return I am practically under siege and in addition my health is not enviable. But I can assure you that I won't stay here long because I can't stand the rushing around, and you can count on it that we will soon be together again and having a quiet chat. My little toad doesn't have much to converse about, but there is something so dear and good about you that the old man is very indulgent and overlooks quite a few faults.

I find the present estimate of the contractor much more reasonable than the earlier one.

---

194 Point of reference unclear.

I don't know what the practice is in Vienna. Do we not need in addition an architect who manages the construction and checks that everything is done conscientiously and correctly? That is how it would be done here. Also, one must always ask for a discount. I am sure he would reduce his bill by 10%. As far as the façade of the house is concerned, I think we should replace the old glass with mirror-glass. The cost is insignificant and it looks good.

Could you find out without being too childish and report to me at once while I'm here? I will then immediately arrange everything with the architect. Dragging it out serves no purpose.

It could turn into a very nice little house – you'll see.

But now: don't worry, be really good, take care of Bella and your little stomach and look hopefully into the future. I myself can't do that, but I wish it for you with all my heart, and send you fondest greetings,

Your sincerely devoted Alfred

135

15 October [1888]

Dear Sofie,

My health was in very bad shape all the time and I had to spend many days in bed. In addition I am so plagued with all those innovations, which don't always proceed smoothly.

You have a little, indeed a big, enemy called Sofie, who does you more harm than you believe. Instead of giving me a clear account to restore to some extent my badly shaken trust in you, you want to tell me that one little woman who spends 100 florins a day must suffer hunger pangs so that I can almost hear the bones rattle all the way from here. You yourself lived on 1 florin per day when your stomach needed more, and you lived well at that time, for you were rosy-cheeked then and you still don't look like a starving woman. You can see, my dear child, that the holes into which you drop your money are unnaturally large, and I have no intention of filling that ocean with my savings.

You have not answered my question how it is possible at all to spend 100 florins a day in Ischl when you don't even have to pay rent. You can't tell me that you are unable to balance income against expenses, as I and a thousand other people do. No one will believe you.

Indeed you have a strange idea of money. If you knew what trouble and care 9/10 of people take to fight for their miserable daily bread, you would soon realize how unfair it is to throw away so much money in such a stupid manner.

And how are you otherwise? How is your health and your mood? I am deeply depressed since my return. I see no one with the exception of those who impose on me for business reasons, and I associate with no one. Tonight is the first time I will visit the theatre, but on my own, and because I am always alone nothing can distract me. If there were someone who could lift my depression even for a few days, I would enjoy going back to work again.

As you know I hate everything connected with spying and investigations. Nevertheless I found it necessary to lift the cover a little because of the endless nasty gossip and the disgusting insinuations. I am pleased to tell you that they say nothing bad about you here in Paris[195] and that all the ugly stories originated in Vienna, Carlsbad, etc. and have been spread everywhere from there. Do you like Vienna and the Viennese so much because you are grateful for that?

Write how you are at present, what you do and how things are. I long for peace and want to leave here and go far away, but in a few days I expect a number of people who will importune me about my innovations and my badly rattled constitution will have to undergo a hard ordeal.

I enclose cash for you. This time it's grace over justice, but you must account for your income and outlay, and I don't want to listen to any more excuses in that respect. So pick up the pen and don't make a fuss.[196]

With sincere greetings from your very hurt but not hard-hearted old

Alfred

There are three bills. Make sure you don't make a mistake and spend them as one!!!!

---

195 Sjöman, *Mitt hjärtebarn*, p. 33, cites an unsigned and undated report from an investigator who confirms that Sofie Hess lived quietly at the apartment on Avenue d'Eylau and had no visitors apart from Nobel.

196 Nobel uses Viennese dialect: *nix muxen.*

136

25 October [1888]

My dear Sofferl,

I wrote to you from the city today and add a few lines to send you cash. Tomorrow I will write in more detail about Döbling[197] if I can possibly find the time. You have no idea how rushed I am here. I suppose that is the real reason for my poor health. For more than four weeks a visit to the dentist is on my to-do list, which is extraordinarily needed, but it is impossible to find a free hour to do it. Paris is an absolute hell for me, because people journey here from all directions and since they usually indicate the time of their arrival in writing, I can't escape them. In addition I have all the technical work, the many letters, the accounts, the construction. It is enough to drive me crazy. Be glad you are a little toad and have no worries. Heartfelt greetings from your old

Alfred

137

26 October [1888]

My dear Sofferl,

It is only now – this morning – that I am able to return your plan with my comments. There is an awful lot of work here. In any case the plan would not reach you earlier, since Sunday intervenes.

I will send your suitcase to your Papa's address.

The money for the villa is ready, but before it is paid out through the Anglo-Oest[reichische] Bank, I want to know that everything is completely in order so that there won't be disputes later on. Tomorrow I will consider everything and write to you. I am under terrible time pressure.

As far as the changes to the villa are concerned, there is still a little time since it will not be vacated immediately. One should never do any

---

197 I.e., concerning the repairs and alterations that would have to be done to the property he is about to buy for Sofie.

construction without a comprehensive estimate and even then difficulties can mount up.

I give you one of many examples. As you know there are Faience tiles in my winter garden. When the garden was enlarged, it turned out that that mould for the Faience tiles no longer exists and I was forced not only to take out the old tiles but rip out and renew the whole cement foundation for the tiles.

It is a great bother. It is a wise rule to change as little as possible, as you will discover.

In my case, that little change will be completed God knows when!

Many heartfelt greetings, dear Sofferl,

Your old Alfred

138

Paris, 27 October 1888

My dear Sofferl,

I enclose (1) the proposed sales contract with my changes or rather comments in pencil, which you should discuss with Dr. Elias.[198] In my opinion you should insist that the house be vacated completely by 20 November, or at any rate not allow the money for the purchase to be paid until the premises have been vacated. Consult Dr. Elias. Purchase renders a rental contract invalid, so there is no legal hindrance. (2) A cheque in the amount of 32,573,37 florins in your name because you will have to pay as the buyer. The sum represents the cost of purchase minus the mortgage, that is 31,813.37 plus fees, expenses, etc. of 700 florins, for which an account will be given later.

I wrote on the back of the cheque how you can transfer the money to Dr. Elias. Have your father accompany you. I assume he has more experience than you.

---

198 On 17 October Sofie's father informed Nobel that the lawyer, Dr Elias, had closed the transaction for the purchase of a house. He mentioned a price of 32,000 florins (taking over a mortgage of 10,000 florins at 4 per cent and paying the rest in cash) and fees of 700 florins. A letter of 25 October suggests that the deal was not in fact closed but awaited Nobel's approval ("I beg your pardon, I meant after your inspection of the contract").

I don't know whether Dr. Elias is reliable. If it is not the usual practice, you could deposit the money with a notary. And that settles the question of money.

You must insist on the house being vacated by 20 November, even if you have to pay three months' rent to the party upstairs, or rather to Mrs Strakosch.[199] It is often difficult to get rid of renters, and one ought to try to avoid such problems.

As for your health: My dear little child, you must try to take things easy. It wouldn't be repentance that makes you ill? That wouldn't be surprising although they say that such a feeling doesn't exist among Jewish women. And in spite of everything you complain of being short of money and being poor – that's ridiculous. Make a little comparison between your outlay since […] and you will have peculiar thoughts about need and surfeit. I don't want you to suffer poverty, but neither do I want you to throw a whole fortune away in this unreasonable manner. Every businessman is building on uncertain ground and when we meet I will show you how my comments are very pertinent just now.

If I wasn't so sick I would immediately go to Petersburg, where affairs make my presence more necessary than ever.

With heartfelt greetings I remain,

Your Alfred

The two marks on the cheque signify that it can be presented for payment only by a bank. One does this with larger amounts for security reasons.

139

6 November [1888]

My dear Sofferl,

So you "gonna be"[200] the owner of a house. But don't start shaking things up until you have asked me about everything, and I have given my consent. Renovating a building is a special matter and requires a great deal of forethought or one runs not only into expenses but also problems.

---

199 The owner of the house? In his letter of 17 October, Sofie's father asks Nobel to remit the price of the house either to Dr Elias or to the son-in-law of Mrs Strakosch.
200· Nobel uses Viennese dialect.

I would already have left if my innovations didn't hold me back by force. In London I was unsuccessful. It would take too long to tell you everything. There was a little accident that could have had worse consequences than it did. It was caused by the stupid safety procedures there.

I returned only yesterday and had an adventure here too. I was in an accident yesterday with the Russian horses. One of them was gravely injured. This was the result of the driver neglecting to give them a run during my absence, and their hot blood was raised. I suffered no harm at all, but it is strange that I was in danger two days in a row.

I have an awful lot to do, and you must not be angry with me when I only manage to put together a few lines, which I send you with heart-felt greetings and wishes.

Your Alfred

140

7 November [1888]

My dear Sofferl,

I haven't had a letter from you in a long time and am beginning to get worried. Are you very careful and guarding yourself well against colds? I will come to Vienna or meet you somewhere south of it as soon as I am done with that cursed gunpowder business. These matters torment me day and night, and you cannot imagine how much I need true peace. I imagine that you will be very comfortable in your little house and will become even more so, but don't think about modernizing. It wouldn't become prettier, only less homely and comfortable.

Don't let anyone mislead you in this manner and present everything to me exactly. Depend on my views, which will certainly be correct in that point.

I have so much work that my head is spinning. If I sleep at night, I can't possibly finish all my work. If I don't sleep I am completely run down. But the way things are going now, I hope to get away soon. The tests at any rate are going quite well.

I take it you have received your suitcase.

My life is often very lonely and sad. I lived for a few days in a hotel, and although life in the house is rather unsafe, I prefer it because all my papers are there. I also do it to avoid expenses.

Both horses are in bad shape after yesterday's accident. It is really troublesome to have your own horses.

I greet you fondly, my dear Sofferl, and wish for your quick recovery and a joyful reunion.

Alfred

Meanwhile I have received your little letter of 4 November.

141

My dear good Sofferl,

You need not send a telegraph because of the four months and the increase of 50 florins. You could take that upon yourself. I very much want you to have a private apartment in Vienna, for several important reasons, which I will explain to you in person, in case you don't understand it yourself.

The old man is ailing for some time now and very run down. It happens every time he strains his mind too much. But what can I do? Matters have to be expedited. A mountain of letters require answers, people must be received, my accounts kept in order, and every day from morning to evening I am in Sévran. I often ask myself how to combine all that. If I engage a secretary there may be an improvement in half a year, but until then there will be more rather than less work for me. Can you imagine the man recommended to me recently as a private secretary was for many years a special envoy and ambassador?[201] Now I'm just waiting to find a hereditary prince for that purpose.

---

201 Gregory Aristarchis (also known as Aristarchi Bey), Ottoman envoy in Washington 1873–83. Later he served as an envoy in the Netherlands, where he died in 1915. Nobel employed him to oblige Carl Lewenhaupt, the Swedish ambassador, who had recommended him. He had doubts, however, about Aristarchi's usefulness as a secretary: "It is difficult to say how far I may succeed in finding a suitable sphere for the regular employment of Monsieur Aristarchi's talents." He eventually employed him to look into the question of promoting peace, but could not see eye to eye with him. He terminated the employment in 1893, telling Lewenhaupt that "he advised me to start a special paper for peace propaganda, to which I replied that I might just as well throw my money out the window" (quoted in Ragnar Sohlman, *Nobel: Dynamite and Peace* [New York, 1929], pp. 227, 235).

If I can send your furniture, to what address should I send it? Don't forget to tell me immediately. It would be best to send it to Döbling, Hirschgasse. But what number?

The time has come when I long for peace so intensely that I think it would be desirable to exchange my life for that of a labourer, just to rid myself of some headaches. My life is awful. I socialize with no one and in any case have no time for socializing.

You ask how Olga is doing. I don't know. She wrote once, but I had no time to answer and I haven't heard anything since. I imagine she will have to be content with small roles, like Hermine,[202] especially because of her wretched health. In any case, aspiring actresses don't rest on a bed of roses.

And now I end my letter, my little cute toad, for every line is an effort. My brain is in bad shape. I wish you all the best in your life and most of all a comfortable home.

Your Alfred

I paid the bills at Louvre's, Bigot's, and Felix's.

142

23 November [1888]

My dear little Sofferl,

I don't know when this torment will end. Every day is worse, so that I begin to curse my life here. And yet I am stuck so fast that I can't see how I can be rescued.

To be candid, I long to see my little cute toad again, and it wouldn't be asking too much if I indulged in another short holiday now. You can't imagine how wretched my life is here. How and why, I must tell you in person.

---

202 Sofie's half-sister, an actress. Sofie's father thanks Nobel for his support for Hermine in a letter of 25 October 1888. For the Hess family see note 192.

With the most friendly greetings from my whole heart,

Your Alfred

On Monday I must once again be in London, and there is so much to do that I am forced to travel through the night. The old man is really to be pitied.

143

Sunday, 25 November [1888][203]

My dear little Sofferl,

It seems to me that that story of Schroll has ghosted around a long time. If it is so good and valuable, I can't believe they wouldn't find a buyer.

You know that I would wish with all my heart a small, nice, but practical home for you. Just don't make the big mistake typical of your fellow countrymen – to admire only appearances. Rather arrange for yourself a truly nice and comfortable life and benefit from the fact that you have met someone who grants it and does not begrudge it to you. And that is a miracle considering how he has been rewarded for it.

Thinking is not the strong suit of my little toad, or you would have discovered immediately that your proposal not to send the furniture to you personally can't work. One must attach a confirmation from the embassy concerning a change of residence, or you would have to pay huge taxes. The cost of transport will be very high in itself, even without the addition of customs duties.

I must go to London tonight. It is very unpleasant to travel through the night, but I couldn't manage it any other way. I had to take the 6 a.m. train to Sévran today to prepare everything there. And now it is likely that a decision will be made in a few days in London, whether I may go on holidays after all those stressful undertakings – a holiday that is now much more important to me than my daily bread. I also want to arrange something concerning your little house, but everything in my schedule has become so tight recently that I was absolutely at a loss how to do things and where to turn.

---

203 The date "Sunday, 25 November" does not work for 1888. More likely, the letter should be dated 1887. The contents suggest that Nobel is still searching for a suitable house for Sofie. His reference to Sofie's furniture also suggests 1887, the year Nobel decided to cancel the lease on her apartment in Paris (see Letter 110).

And now, my dear Sofferl, my time is up. I am about to drive to the train station but before I go, I send you heartfelt greetings.

Your Alfred

My address in London is the Grand Hotel, but only for one or two days, unless something unforeseen happens.

144

9 December [1888]

Dear Sofie,

Your short letter which arrived yesterday sounds so sad and down-hearted that it moved me. I am not as hard as a rock, and although I had reason to be bitterly angry, I won't grumble today.

As far as I am concerned, I suffer only one evil, which is worse than other serious ailments – insomnia. Sometimes a week goes by without my closing my eyes and I have become indescribably nervous. I enclose one of those pieces of paper that are hard to come by, whereas you seem to think that they, like manna, fall from heaven and cost nothing.

With heartfelt greetings,

Your old Alfred

145

13 December [1888]

Dear Sofie,

I have no news to tell except that I have a bad cold. The whole world seems to be sick this year, however, and so it would have been strange if I had been spared. Yet I must soon travel to London, sick or not.

You don't need to be embarrassed about feeling deeply sorry for your poor little doggy Bella. On the contrary, that sort of thing always touches me.

Best,

Alfred

146

[1888/9]

Dear Sofie,

A person who knows no consideration must learn consideration. Life is governed by iron laws. Anyone who believes they owe no consideration to anyone yet demands every consideration from others will certainly be disappointed. They say Christ's benevolence was unlimited – if you believe it – but that doesn't work for the rest of us people. It almost looks as if you thought my good will and forbearance are based on stupidity, but you are very wrong in that point. That point of view, by the way, seems quite natural, for in my experience Israelites never do anything out of good will. They act merely out of selfishness or a desire to show off, and how can they understand a trait in another person that they absolutely lack themselves? The Israelites have some very good traits, which I always acknowledge, but among selfish and inconsiderate people, they are the most selfish and inconsiderate. For them it is "self and family" – all others exist only to be fleeced. Perhaps they are right to act thus, but then they shouldn't be surprised if they are treated as they treat or want to treat others.

Let's assume I was poor and your family rich – do you think they would lend me even a penny without interest and compounded interest? If you believe that, you don't know your Pappenheimer,[204] as I know them.

You yourself appear to be unselfish, but in essence your generosity and your wastefulness are based more on unreason than true unselfishness. You just don't understand the value of money until you are penniless, and yet you can be quite small-minded and fight over every penny. To err is human, so let's not talk about it, but to damage the honour of an innocent man is many steps beneath error, and does not admit any excuse.

The registered letter I sent yesterday should have reached you in the meantime.

Many sincere greetings,

Alfred

---

204 Quoting Schiller's drama *Wallenstein*. The expression became proverbial, meaning "I know this type very well."

147

[1888/89]

Dear Sofie,

I am sorry to see and truly pity you when I see from your letter and
your telegrams that you have still not decided on an apartment.[205] And
always just because you think nothing is good enough for you. And no
one will be able to help you because you are always so discontented
and want more. I do understand that you may not be able to come to
full terms with your position in life, but there is no reason to make
things worse. To live in disgusting furnished quarters for years does
not contribute to comfort in life, and once you have a fixed residence
and a true home, you will also find your life more bearable.

And one more thing, dear Sofie, learn to be considerate to people
who are not related to you, for example, to me. You would not believe
how much you could profit in life if you understood what one owes to
others.

Also consult with your sister Bertha,[206] who means well, and obtain
at last a comfortable home, not too large, because that tends to become
burdensome, and that's the end of any comfort.

I feel very, very weak at times and strongly sense that my days are
waning. Use the time therefore, before I embark on the shortest and yet
the farthest journey and arrange your life such as I wish it.

Your benevolent, devoted Alfred who pities you

148

9 March [1889]

Dear Sofie,

My work is never-ending. I am bothered and tormented, and have
no time to answer your last letter. There are so many unpleasant and

---

205 A letter from Sofie Hess's father to Nobel (19 August 1889) indicates that she was
     looking for an apartment in Vienna. Since he asks what price range would be
     acceptable to Nobel, it appears that Nobel was willing to finance it. See Letter 150.
206 Bertha Goldmann, née Hess. By 1892 she and her family had emigrated to America
     (see SH 14). For the Hess family see Letter 132 note.

troublesome things going on here[207] that I am losing my mind. And in addition my health is extremely poor. I suffer from heart cramps almost every second day, and that makes my life difficult.

Heartfelt greetings,

Alfred

149

13 April [1889]

Dear Sofie,

I enclose two Hungarians, worth about 4,700 francs, that is, about 2,150 florins. And I return the statement sent to me, because it contains a receipt, which you must certainly keep safe.

As I can see from your letter you are not very pleased with your neighbours, perhaps because you don't know how to handle them. In any case no one will suffer the kind of treatment I bore on account of my good heart. It seems you considered that a weakness in me – that is a great error on your part.

Sincere greetings,

Alfred

150

14 May, [1889]

My dear Sofferl,

You have not answered my question whether one must hire an architect in addition to a contractor in Vienna, as one does elsewhere for every construction. I suppose that cannot easily be avoided.

---

207 A first reference to the difficulties leading up to Nobel's sale of his patent for ballistite to Italy (see Letter 155 and note) and the financial scandal involving Nobel's partner Paul Barbe (see Letters 165 and 171).

If you don't like the house in Döbling, I don't see why it has been bought. I would find people who would like it very much and would take pleasure to live in it, among others yours truly.

If you don't want to live in the house, rent an apartment and furnish it, and get the idea of a palais out of your head. I myself have enough trouble with my cursed house without helping other people to similar difficulties. And then – one might buy the whole world if one didn't have to pay for it. But money isn't water, and even the latter must be paid for in the city.

I will ask at Bigot's about the buffet mirror. It would be unprecedented to find that it has disappeared.

I enclose a letter from Gschwandtner,[208] which we discussed already in Vienna. Something has to be done about it. In any case, it is not only expensive but also inconvenient to have residences in all corners of the world. I have my residence in France and don't see why I should have another residence in Austria.

If I grumble a bit, it is because I have so much trouble and am sick as well. I would need a whole year of complete peace to recover, and to practice science purely as a hobby rather than a business. Only the gods know when such a golden era will dawn. People are again besieging me with telegraphic summons to London, to Scotland, Hamburg, Italy, etc. What a life! Here everything is green and very beautiful, but I have no time and no mind for anything.

With heartfelt greetings,

Alfred

151

[1889]

My dear Sofferl,

Your views have always been a little screwed up, and they don't seem to improve. How would such a little, dependent, barely educated creature without talents look in a palais? At best you would be a laughing stock

---

208 On Gschwandtner see Letter 68.

for others. And how much effort and trouble it is to keep a palais in order, not to speak of the great and unnecessary costs it causes and that in the long run only bring about endless problems. Little birds, however fat, belong in little cages and will feel most comfortable there.[209]

So much for the argument from reason and comfort. But there is another, more serious argument. How do you imagine going about buying houses? Do you think the owners will be glad to give their possessions away as a gift? Not even Rothschild can buy whatever he sees, and one must understand the value of money after the collapse of the copper market. Don't let yourself be persuaded by the Jews there that I am so rich – you could be in for a great disappointment.

It is a strange business, my dear little toad. You are looking everywhere for a palais and none seems good enough for you, and in the meantime you live year after year in disgusting furnished quarters and on account of your castle-mania lack a comfortable home. Really, you don't know what you want. You lack something that the most luxurious walls can't replace. Explaining that to you is all very well, but you keep returning to your childish views.

Look, dear little child, if you really could have such a palais, everything else would have to be furnished accordingly and there would be much unpleasantness, given the position of your family and other conditions that I pass over in silence. Believe me, if you can't be happy with a simple life, you will not achieve happiness with a pretentious residence. I don't know what dough you are made of, that you can't learn simplicity from me. Imagine how you would have turned out in the hands of a pompous person. The street urchins would have made fun of your appearance. Be grateful therefore that I pointed you to a better course and make a nice curtsy.

Many heartfelt greetings,

Your Alfred

209 See Letter 147.

152

Zurich, 27 August [1889]

Dear Sofie,

Fall is coming, and you still don't have a home. Really, you are a little goose. And always because you think you must live in a palace, yet you are living for years now in miserably furnished, uncomfortable, dirty quarters.[210] After all, Vienna is large enough to find a home, and your stories seem to me silly and implausible. Hurry up and establish yourself comfortably while I am still alive and can help you.

I enclose a bit of blue, but don't treat it like blue smoke. I don't know where I'll go. Peace is my desire.

Heartfelt greetings,

Alfred

How is your sister Bertha?[211]

153

1 September [1889]

You sent me a telegram that you want to come to Zurich. To be honest I have tasted enough of your hurtful behaviour and want to live what's left of my life as peacefully as possible. In any case, it would not suit you to come here. In this hotel alone there are eight people

---

210 It appears that Sofie continued to live in rented quarters in Vienna and never moved into the villa in Döbling, perhaps because Nobel refused to finance the renovations she wanted (see Letter 156) or because it was too far from the centre of the city for her taste. Between 1890 and 1892 she rented quarters that Nobel agreed were arranged "nicely and comfortably" but were too "palatial" for her station in life (see Letters 164, 182, 199, 207). Perhaps she rented the villa in Döbling to a third party, but there is no further mention of it in the rest of the correspondence. In view of her irresponsible handling of finances (see Introduction, p. 15), it is likely that she borrowed money against the property and was forced to sell it. In 1893 she is once again on the lookout for an apartment (SH 16) and after an unsuccessful search moves into a hotel (SH 17).

211 Heinrich Hess thanked Nobel in a letter of 19 August 1889 for helping Bertha. The nature of his help (presumably financial) is not specified.

with contagious and life-threatening illnesses. I knew that, but I came anyway because I don't care much for my old bones, and no one would miss me. Not even Bella would shed a tear, and I believe she would be the most sincere creature in her grief. At least she would not, like all others, look for any gold pieces bequeathed to her. By the way, those dear people will be strangely disappointed, and I enjoy in advance the thought of their eyes opening wide and the many curses they will utter when they see there is no money.

How is Hebentanz #2?[212] At least you can't bring us together since I avoid not only you but the country. One deceit less – you will be so sorry, won't you? It was always so nice and comfortable to present me to the gentlemen and afterwards make fun of me with them. Of course you didn't explain to anyone that I acted the way I did only out of generosity. No, that would have been too bad and spoiled the effect.

But go on with your deceit. Perhaps that is the right way to acknowledge the good deeds from which you benefited. Soon no one will be able to hurt me because I live so alone that even gossip passes me by without my taking notice. In any case I am so close to the grave that even the stings of injured honour become blunted.

Many sincere greetings,

Alfred
St. Moritz Bad, Engadin, Hotel du Lac

154

St. Moritz, 4 September [1889]

Dear Sofie,

If I write rather hard words, don't think that I am angry with you. I'm not angry with anyone, I merely attempt to speak clearly to make you understand at last how gravely you have sinned against me. But it seems to me that it is impossible to make clear to you what honour means and that one must not step on it – less even than one must step on corns. In any case I have been nicely rewarded for my long years of kindness and forbearance.

---

212 I.e., another deceitful admirer like Dr H[ebentanz]. See Letter 111 and note 80.

I quite believe that you have worries, but only worries about your-self, not about what you did to others.

The fact that you always try to introduce me to people with whom you start a relationship points to such a low character that I sometimes regret what I have done for you. And I have to take that from a person who once accused me of being brutal. Look into the mirror and you will wonder about all your brutalities against me. My brutality seems to consist of helping people who don't deserve it.

You can always address your letters to me in Paris. One was forward-ed to me today.

From here I probably have to go once again to Turin and then to Stockholm. Difficult, cold times: storms at sea and storms in my life.

Many sincere greetings,

Alfred

I am glad to hear that Frieda[213] works out well. Don't teach her the way you taught Olga. Apropos Olga: they tell me she was holed up in Fran-zensbad for almost two months. I suppose she is sick – that is the re-venge of sins of the youth.

155

Lucerne, 21 September [1889]

Dear Sofie,

I am en route from Turin[214] to Stockholm – my health is completely bro-ken down, but this time it is even more necessary to go there because it is the last time I make that journey.

You have treated me badly, very badly, in return for my goodness and forbearance, but I won't take revenge and would sincerely like to see you have a comfortable home and a life without worries.

---

213 A hired companion or domestic. Heinrich Hess reported in August that Sofie dismissed her long-time domestic Marie for being insulting and for quarreling with the cook. For Olga, see Letter 33 and note.

214 On 16 September Nobel signed over his patent for ballistite (smokeless gunpowder) to the Italian government in exchange for licencing fees and a large order for his factory in Avigliana.

I have recently suffered once again serious losses, and who knows what the future will bring. But that's all the same to me now. Can anything worse happen to me than has already happened?

Heartfelt greetings,

Alfred

156

[1889]

Dear Sofie,

I just received your little letter. It is your fault that you are stuck as far as finding a home is concerned. You decided to arrange for a smoking room and other rooms designed for your gentlemen lovers, at my cost. Obviously you think I am too good or too stupid to mind, probably the latter. And yet you comment in every letter about losing weight on account of worries, but an acquaintance of yours whom I met in the street yesterday assured me that you never looked better. Therefore your laments have no other purpose than to deceive me. But you won't do that much longer, because I know my Pappenheimer, as the proverb has it.[215]

And yet I can't but pity you. Don't you have a trace of conscience that tells you occasionally how shamefully you rewarded my care? And now you are stuck, isolated and without friends and any true family bonds, and soon also without youth, which covers some shadows and beautifies them. Why did you resort to that ugly lie that covers everything up instead of talking openly to me, a man who means well towards everyone. I bring no other accusation against you but your endless ugly lying and that you always found the cleverest and most devilish ways to expose me to ridicule. There can be no doubt what kind of relationship existed or exists between you and [...],[216] especially now that I have your written confession in my hands. And even though the domestics knew of it, you always tried to set up meetings between him and me, and in such a way that courtesy did not allow me

---

215 See note 203.
216 Illegible.

to withdraw. The more I think about it, the meaner I find the mind that can conceive of such a thing.

I am now certain that much about which I was earlier in doubt is unfortunately true and that my goodness and forbearance were abused in such a way that any punishment, however hard, would still be mild. But I want to forget, forget, and forget again. If I could only do it, I would flee to the end of the world, but even there lie shadows of the past. Only the final sleep will cleanse me of all the deceit with which my innocence has been stained.

The whole tragedy came about because of your helplessness. Who can bear to thrust out into the world such a weak creature, as long as one believes that she is truly orphaned?

Heartfelt greetings,

Alfred

157

11 November [1889]

Dear Sofie,

I quite believe that your life is rather empty and joyless, and I sincerely and heartily pity you in that respect. But what do you expect when you deliberately hurt and abuse the whole world and do not curb your own whims and misbehaviour?

You acted against me in such a way that I must be the only person on earth who would not abandon you completely to your fate in these circumstances. You have so tarnished my name that I, who do nothing but work and help others, am discredited and must live an isolated life. You say I am able to live alone and be content and happy. But that is unfortunately not the case. Illness and fatigue are wearing me down and often, before I fall asleep, I think of the sad end I will meet one day, in the company of a single old domestic who wonders all day whether I have bequeathed anything to him. He does not know that I will not leave a will – I have torn up the one I made once – and that my finances are more and more in tatters. Any man who throws away money the way I did can't have much left. I have engaged in financial speculations to drive away my dark thoughts, and now I have suffered huge losses. But it's all the same to me, and if my old ragged heart still bleeds, it is only over the profound shame others brought on me – a man who has always been generous and acted nobly.

I will send the money you owe to Hortense.[217]

Your fantastic stories about furnishing a room for gentlemen surpass everything that has happened so far. Do you seriously believe that I would allow you to accommodate the young man at my cost? In any case there can be nothing more stupid than for you to live in Vienna. You have embarrassed me and yourself there. Every stone can tell a story. That doesn't touch you because you have no sense of honour, not the slightest idea of it, but it can't be pleasant to have such a reputation and live where one is so much exposed to gossip.

Recently, when I called on a gentleman from Petersburg at his hotel, I ran into an acquaintance from Vienna who told me some nice stories. People are nasty gossips, I know, because there was a woman there too, but even nasty people can't find material if there is none. From what I hear, you are again frequently in the company of your admirer. I would have no objections if my honest name wasn't involved.

My health is very, very poor. The instances of fainting have become less frequent, but I am sick in body and spirit. Nasty people have poisoned my life.

Many sincere greetings,

Alfred

158

[November 1889]

Dear Sofie,

There *are* things that are new under the sun,[218] because there has not been a simpleton like you since the beginning of the world. You still don't understand that I have no intention to allow you to swallow up immense sums every year and spend them in the stupidest way. Why? What for? Should I allow it out of gratitude for the insults you made me suffer and the abysmal, hair-raising lack of consideration you showed?

---

217 Not identified.
218 An allusion to Eccl. 1:9: "There is nothing new under the sun."

[…] You and your father and all the others seem to make a joint effort to discover how far they can go before my patience runs out. You want to find out? You will!

It is absolute hair-raising. Since 1 July, that is, in less than six months, you have wasted 48,267 francs.[219] What stupidity and what a sin!

Alfred

159

[1889]

Dear Sofie,

I return the enclosed estimate that you sent me, but don't get my name mixed up with it, because I have ordered nothing and have not asked for an estimate. In any case I am ill and need to be left alone rather than being reminded how you abused my name.

You are too naïve to understand that a man is attached to the purity of his reputation. In that respect you will never see the light. And that is the only mitigating circumstance in judging your actions against me. Yet, I think it is rare in life that a person is rewarded like this for his goodness and generosity.

But now I want to draw the curtain of oblivion and forgiveness over the past, as far as I can. I am no good at exacting a penance. I know I should do it more thoroughly, but I can't. Why? Because I pity you and because you carry the heavy burden of your own transgressions. How is it possible to break down even my forbearance? One needs true skill to do that, and in that respect you are a true artist. But you have an empty and sad life, and truly you have a certain charm and pleasant personality that deserved a better fate. But your family lacks any sense

---

219 Sofie's father wrote to Nobel on 1 December (HH 15 in Arkiv Ö I-5), expressing his astonishment at the sum mentioned and asking him to forgive her and "cast the mantle of oblivion over it." Apparently Nobel had asked him what he considered a reasonable amount for Sofie to spend on living comfortably, but Hess declined to make any suggestions. In another letter, of 14 December (HH 16 in Arkiv Ö I-5), he asks Nobel to give Sofie another chance and resume living with her. "You will find that she has become more serious," he writes. He cannot believe that Nobel "does not care whether or not she has admirers" and insists that Sofie "loves him as deeply as ever."

of honour, and that is the root of all the evil. Without a sense of honour one can't get along with an honest person, who feels hurt and wants to stand before the world unsullied. The result is rather sad for you. I assure you of my true and heartfelt pity.

Your devoted Alfred

There can be no talk about a room for gentlemen.

160

Berlin, 22 December [1889]

Dear Sofie

I am sick and worried and don't want to associate with anyone. Least of all do I want to meet with you at this time, for that would refresh the memories of all the damage you have done to my honour and how you have rewarded my goodness. And now my heart is so sick and hurt that it needs twice as much ease. I say "sick" because a few hours after my arrival here, I suffered the old evil, a heart cramp, such as I have not had in a long time. That is a warning of what will inevitably come, and I want to get home as soon as possible to put everything in order as long as there is time. There are many loose ends, however. Unfortunately, I am stuck here because a part of my luggage that contains important papers was left behind. But they assure me that I will soon receive them.

Make the best of the Christmas celebrations. You have many relatives, and their children all love you. But you spoil your life being hateful and don't realize that very often the ingratitude and lack of attention you criticize in others is a response to your own behaviour. You are quite good and generous, especially with other people's money, but you give in a patronizing manner that leaves a bitter taste behind and you are absolutely incapable of self-criticism or of discovering that not everything you do is perfect. I don't say that to reproach you but to open your eyes and to make your relatives become as fond of you as you essentially need and perhaps merit on their part.

Where do you stay now and how have you managed with the domestics? I hope you have some peace and have found servants with the skin of a rhinoceros who will quietly accept that you treat them like dogs, and I don't mean, like Bella.

Gather the children around you at Christmas and decorate a pretty tree so that you will find joy in the joy of the little ones. That is what I wish you from my heart

Your old Afred

Olga is in Berlin. In other circumstances I would quite like to see her, but at present I have only one wish – to be alone, completely alone, and I am alone except for the unavoidable waiter and the man who looks after the stoves.

161

December [1889]

Dear Sofie,

I was very sorry to hear that you too are sick, and I want to warn you to be as careful as possible during the convalescence. For the illness could return, and that would certainly be no fun. Here more than a million people are sick, and the mortality rate is worse than during the worst cholera epidemic.

I fell ill in Berlin, but nevertheless did not want to delay my journey and therefore could not get to Cologne, and had to stay, sick as I was, in a miserable hotel in Dortmund.

Worse than influenza was the heart cramp that has lately become a daily occurrence and makes for painful, sleepless nights. It is a dire illness and an ugly harbinger of things to come.

Write how your life is arranged now, dear Sofie. I quite like Vienna and would rather like to go there if you had not made life for me impossible there. A clean reputation is more necessary to me than a clean shirt, because a shirt can be laundered, but a reputation can't, and certainly if anyone has earned a clean reputation it is me. I never thought of myself. I always and only cared for others.

Sincere greetings to your family. I wish them a lucky and happy New Year. To write individual letters I lack both time and courage. My strength is broken, and quiet leisure is my only pleasure. I wish you good health, consideration for your fellow human beings, suppression of all silly hatred, the joy and peace of a home. Don't always blame others for what you have caused yourself, and you may yet find good and true friends. So far your own lack of consideration was your worst enemy. I say the truth in all sincerity and goodwill,

Your much devoted Alfred

Saturday

161a

January [1890]

Dear Sofie,

Today I am out of bed for the first time and am able to reply to your letter, though briefly. I am very glad to hear you finally have a nice home. You have my best wishes for your well-being there and for your contentment – that is true wealth. I would like to visit you there, but found during our stay on the Semmering that our views about propriety and decency are further apart than ever, and since I am not in the mood to be treated like a stupid boy in the long run, our relationship has come to an end, I think. If what you call *Ungeniertheit*[220] will benefit you, I won't analyse it further. If I was your dependent, I would not stand it even for two days, but as your benefactor I must put up with quite a bit. But even my forbearance has limits if my daily and hourly reward is to be ridiculed. Well, gone is gone. Let's leave it at that.

Many thanks for the beautiful buttons, which are very elegant and nice. I had too many worries last year to think about Christmas presents, and my broken health did not always allow me to bother with shopping. I did not send you any presents and I think people will also excuse me on account of being in mourning.[221]

Tomorrow I may perhaps be well enough to do something about your suitcases and Bigot's boxes. But now I am so tired that the pen is practically falling out of my hand. I can only send you heartfelt greetings and wish you all the best.

Your devoted Alfred

Monday

What is your address now?

---

220 I.e., nonchalance.
221 For his mother, who died in December 1889.

162

27 February [1890]

Dear Sofie,

I still can't get away. These cursed innovations are as sticky as pitch, and every day brings with it something unforeseen that must be looked after. If I had known what torment would arise from this project, I would certainly not have started it. If I departed now, I would have to charge back immediately, whereas I wish for a longer period of quiet.

Day after tomorrow is your birthday, and I would have liked to give you my wishes in person. But since it can't be done, this letter must express my heartfelt wishes to you.

I am not one to bear a grudge and I forgive you gladly whatever can be forgiven. A man who is angry at others spoils only his own mood and perhaps brings most harm on himself. But what is due to one's honour and dignity is a different matter. Before I go to Vienna, I want to know and be very clear about one thing: Have you embarrassed yourself and indirectly me? Perhaps you don't understand me, for in your family the sense of honour is very poorly developed and one makes fun of decency. But I am not forced to accept other people's standards and so I keep to my belief and do not want either to be evil or be made out as evil by others. But now I am very much afraid that you have started a relationship of the kind that makes my presence in Vienna impossible. Let us understand each other: I mean I can't live at your place without exposing myself to sneers and ridicule, and I don't want to stay elsewhere and hurt you. That is how things are, and perhaps you have not understood my letters properly when you thought we would see each other in Vienna. I meant we could arrange for a meeting elsewhere, and we could then talk to establish whether a visit to Vienna is at all possible for me.

Don't regard this as spoken in any negative sense. On the contrary, I meant to speak from the heart.

Your very forbearing and very well-meaning Alfred

163

March/April [1890]

Dear Sofie,

If I am to come to Vienna, it is absolutely essential for me to have a suite in a hotel, where I may receive business associates and to which my mail may be addressed. That is a matter of formality but unavoidable and necessary, for more people are searching me out now than ever before, and you will understand well that I can't embarrass myself.

Your health, I understand, is not good, but that is better cured with a reasonable diet than with a stay in Carlsbad. We'll talk about it. I am very sorry for you because you are such a helpless little creature and if you were not so moody and spiteful we would get along better.

Best,

Alfred

164

Paris, April [1890]

Dear Sofferl,

I must honestly tell you that I longed to be away again as soon as I arrived here. I am once more up to my ears in work, and the assistance I get is rather meagre, or else things would progress more quickly. The things I have in train now cannot be left unfinished, or else I would indulge in a little quiet time, for my need for it becomes more noticeable every day. And I must say I enjoyed more quiet during my stay in Vienna than in many years now. You have arranged everything very nicely,[222] I must say, but you would need domestics of a different sort, and that requires more generosity and most importantly more distance, and a more even and dignified treatment of your domestics. But as you know I'm not

---

222 This is also reflected in a follow-up letter from Sofie's father: "As Sofie told me, you expressed satisfaction during your visit concerning the purchase of furniture" (HH 12 of 14 May 1890 in Arkiv Ö I-5). His letter further indicates that they discussed an annuity for Sofie. Her father notes the difference between a *Leibrente* (payment for lifetime) and life insurance, which requires a medical examination. See Letter 164a.

born a sermonizer and of all thankless tasks preaching in the desert is the worst. Today is Easter Sunday and the air is thick with sermons from the pulpit, and for that very reason I want to save my sermon.

You made a big mistake to take such large quarters and put such a little – though fat – bird into an immense cage, but apart from that I must compliment you: you have arranged everything really nicely and comfortably. What's mainly missing are two husbands – one for you and another for Bella. They must pamper you both.

As soon as I have time to breathe I will get a few blue bits of paper or something similar and send it to you, as I promised before my departure.

In the Orient Express I found no one with whom I could exchange even two words. In any case, it seems that travellers nowadays consider each other as wild animals with whom it would be dangerous to make any connection. Luckily I had books with me: *Die Waffen nieder* and *Salammbo*[223] helped me pass the time.

Write and tell me how your cough is faring, or rather not faring, dear Sofferl. I hope you are completely over it, although the weather was not very favourable. If necessary, drive to Baden and get rest a few days in the clean air. I wish you a cheerful, comfortable life from my heart and fond greetings,

Your old Alfred

Sunday

164a

8 April [1890][224]

Respected Mr Hess,

I would like to secure for your daughter an annuity, not to countenance her insane wastefulness but so that she will not suffer poverty in the event of my death. Could you provide me with the necessary

---

223 Bertha von Suttner's pacifist manifesto, published in December 1889, and a historical novel by Gustave Flaubert, published in 1862.

224 The original of this letter (Swedish translation in Sjöman, *Mitt hjärtebarn*, p. 266) was preserved by Sofie's granddaughter Olga Böhm (d. 1995). A copy of the original is among Sjöman's papers in the National Library of Sweden.

information in that respect, that is, which [insurance] companies are the safest and how much interest the invested capital would yield, i.e., what is the rate of return.[225] To determine the conditions, the physician of the insurance company will probably have to confer with your daughter's physician or physicians, who have looked after her and are familiar with the state of her health.[226]

May I ask you to look into this matter?

Yours sincerely,

A. Nobel

Mr Goldmann[227] should be able to give you valuable information in this respect.

165

[Paris, May 1890]

Dear Sofferl,

Don't be surprised that I write so rarely. It may have been my health. I was sick for several days and am still quite wobbly. But that is a small matter by comparison with the insults the papers heap on me now,[228] and the difficulties and troubles I have to contest with. You cannot imagine the chicanery to which I am subjected and how much time and effort is involved. But I hope I will soon get to the point where I can depart and someone will joyfully turn his back on Paris – that one is me.

It is like summer here, but I couldn't experience it because I haven't been able to leave my room the last four days. I caught a nasty cold. From

---

225 Heinrich Hess replied on 13 April, mentioning three possible companies and their rate of return: "For every 1,000 florins paid in, there would be an immediate return of 60 florins per annum" (HH 13 in Arkiv ÖI-5).

226 Heinrich Hess noted that no physical examination was needed in the circumstances (HH 13 in Arkiv ÖI-5).

227 The husband of Heinrich Hess's daughter Bertha? He was a journalist by profession.

228 The attacks in the wake of the financial scandal involving Paul Barbe, Nobel's partner, and the sale of his patent for ballistite to the Italian government after the French government had passed on the deal. Nobel was publicly accused of work espionage and threatened with imprisonment by the prefect of the Départments Seine et Oise. After his laboratory in Sévran was raided by the police, Nobel went into voluntary exile (see Letter 172b). He spent the last years of his life in San Remo, Italy.

my window I see beautiful greenery and would love to be in the country. In the meantime I send you heartfelt greetings, my little childish toad.
Your old Alfred
Sunday

Certain things will change as a result of what happened here.

166

[Turin], 31 May [1890]

Dear Sofferl,

I am stuck here in Turin now for three days, seriously ill and occupied in a very sad manner. Perhaps I'll make a quick trip from here to Vienna, but I don't have much time. But what I have suffered has aged me a great deal. I not only feel it, I look like an old man.
Best wishes.

Your old Alfred, whose whole soul is out of tune.

167

Dresden, 17 June [1890]

Dear Sofferl,

Dresden is a kind of Venice, but without the noise. No one knows me here and I can finally rest from all the strain. When I say strain, I mean the endless, for me absolutely grueling, gossip of the domestics. No wasp hums so incessantly into one's ears and none has a poison with such long-lasting effects. People in Vienna don't want and cannot understand me, or they would realize how little I ask and how easy it is to fulfil my incredibly modest requests.

You are certainly to be pitied, more or less, and there are certain mitigating circumstances in judging you, but you yourself would be much happier if you could learn to show consideration. Then you would discover and realize how gentle other people can be.

With many heartfelt greetings from the solitude in which I find myself once again.

Your devoted Alfred

168

Berlin, 20 June [1890]

Dear Sofferl,

I dislike it as much here as I liked it in Dresden. One reason is that I've caught a serious cold en route. You can't imagine how awful the weather here is, a regular alliance between fall and winter. My summer clothes were not at all suitable, and immediately I caught cold.

For the time being I have much to do here and the gentlemen from Hamburg announced per telephone that they would visit me tomorrow.

After that I am free to depart, but in this bad and stormy weather I have some hesitations to undertake the long sea journey, especially since Liedbeck[229] is in Norway and will return only next week. So I asked around among my relatives whether they would like to come to Hamburg and meet me there. I am so tired of travelling and especially travelling alone affects me. I would like to have young people around me when I travel to the factories.

And now – how is my little toad? Is your health better? You will see, as soon as the weather warms up, you will feel much, much better, but you must not get angry, for anger is poison. What you call your hot temper is, however, not even anger but just bad manners and not becoming at all.

It looks like your bad manners affect your intestines and you punish yourself, but that doesn't help others.

Perhaps I drive to Hamburg with the gentlemen tomorrow. By the way, I don't know exactly what they want from me. They send telegrams and telephone to say that it is an important matter. In any case I'll send you a telegram once I know myself where I'm going.

All the best,

Your old Alfred, who forgives much but not everything.

How I would like to return to Dresden. I like the city so much. It has always been my favourite holiday spot. The place is pretty and the people friendly.

I have much to do here at my dentist's. It appears that the condition of my gums is dismal. Tomorrow he will cauterize them again, but he

---

229 For Liedbeck see note 2.

doesn't dare to put in fillings because the inflammation is so serious. For some time now, I've often been suffering from a toothache.

169

June [1890]

Dear Sofferl,

Poor me. I really don't have a chance. After I paid my hotel bill this morning and was on the point of departing for Stockholm, they brought me a registered letter from London, which forces me to travel there immediately. It concerns an incredibly audacious scam,[230] which will do great damage to my interests. I will have peace only in my grave, and probably not even there, because I have a presentiment that they will bury me alive.[231]

Many heartfelt greetings,

Your very aggrieved Alfred

170

7 July [1890]

Dear Sofie,

We can't go on with the certain something as before. For some time now my expenses have been much greater than my income, and although I am very generous, I'm not an ass and have no intention of ruining myself for other people. So much by way of preamble.

If you and your family had a little more feeling you would have realized long ago that things can't go on like this.

My life is becoming ever more unpalatable, and evil people make sure that I have no peace. The long journey to England and all the new

---

230 A reference to Frederick Abel's machinations, which ended in a protracted lawsuit over the patent for ballistite in England (the "Cordite Case," begun in 1888). The verdict rendered in 1892 went against Nobel and he was obliged to pay the costs.
231 He therefore gave instructions for his body to be cremated.

problems and lawsuits that have arisen take a lot out of me. It is becoming clearer to me that this strenuous life is affecting my health.

I was going to go to Stockholm, where I must do important work with Liedbeck. As soon as I arrived in Berlin, the runaround with Hamburg began. They sent me a telegram not to depart under any circumstances without awaiting the arrival of the gentlemen to discuss a very important matter. When that was concluded and I was already on the point of buying my tickets, I received a telegram from London that made my trip to Sweden impossible and forced me to depart for London, and that via Paris to pick up important papers there. That is only the beginning of a nasty lawsuit that will result in endless costs and troubles.

If I compare my life with that of other people I must say there can be nothing more disconsolate. And if I act with goodwill and nobility towards others, how thoroughly they abuse me, and how few people understand and respect my actions!

You asked me to buy diverse things for you, but I had a very different business to look after and I have enough going on in my head without running around to do your shopping. I don't even know at this point how to find time to eat.

Heartfelt greetings,

Your old Alfred

I hope you don't have such miserable weather as we have here. I am freezing even in my winter coat.

171

[August 1890]

Dear Sofferl,

This year I have nothing but bad luck. Now the death of Barbe[232] forces me once more to return to Paris. I would already be on my way if I did

---

232 Barbe committed suicide on 29 July 1890, when the financial scandal caught up with him. Nobel took over as chair of his enterprises and reorganized his business after the extent of Barbe's speculations became apparent. The details of his shady dealings came out only in the Panama trial 1892/3. They caused Nobel enormous losses, which required him to take out a bond loan. He dismissed the board of directors and put a new managing director, Paul Du Buit, in charge of his enterprises.

not have to wait for documents that absolutely must be provided before my departure and that I expect with the mail from London.

The death of Barbe will involve me in a lot of difficulties and burden me with a huge amount of work. Who knows what complications may arise from it!

The journey here has affected me so greatly that I have still not recovered and now I am in for another strenuous journey. It is strange what a toll the last few years have taken on me and how little it took to exhaust my energies completely. Please look carefully after your health, embarrass me as little as possible, take care of Bella's asthma, and accept my heartfelt greetings

You old Alfred

By error I wrote on two pages and need to cut them off.

171a[233]

Monday [summer, 1890]

Do you want to go to Carlsbad? If you think it would be beneficial, don't put it off too long, for if the weather turns cold, it could be detrimental for the treatment.

Decide one way or another. I myself would also like to go to Carlsbad, but I can't guarantee that it will be as long as proper treatment requires.

I will write to you today by registered post and for now send only sincere greetings,

Alfred

233 Swedish translation in Sjöman, *Mitt hjärtebarn*, pp. 274–5, from an original in the possession of Sofie Hess's granddaughter, Olga Böhm; a copy of the original is now among Sjöman's papers in the National Library in Stockholm.

172

Wednesday August [1890]

Dear little Sofferl,

Why am I stuck here so long? It's not that I let the summer go by because I am healthy and need no holidays, as you think. On the contrary I am so ill that the doctor thinks it is madness to continue working under such conditions. But I can't help it. The death of Barbe had caused terrible difficulties and led to complications, and it's almost a question of "to be or not to be" for me.

I helped Hermine[234] according to my stupid habit, to that she did not have to bother you about it.

As far as Olga is concerned, I withdrew my support from her a long time ago. That was the result of hearing only very negative reports about the young lady from acquaintances in Berlin. She did not seem to deserve my help. After the report from Lübeck was so favourable, I decided to make the difficult career[235] she had chosen a little easier for her, but it seems that the report from Lübeck was false – my reports from Berlin are from a reliable source.

I hope I can use part of September to look after my health. More about that in a few days.

Many heartfelt greetings,

Your devoted Alfred

I am terribly busy and suffer from a serious stomach catarrh.

172a[236]

Hamburg, 7 August 1890

Dear little Sofferl,

The summer goes by without any rest for me and yet I've never needed rest more urgently. The death in Paris overturned all my plans, and I had much inconvenience from it. If that were not the case, I would not

---

234 Heinrich Hess thanks Nobel for his help in a letter of 21 July (HH 14 in Arkiv Ö I-5).
235 I.e., a career on stage.
236 Unnumbered in Nobels Arkiv ÖI-5.

embark once again on the long journey to Paris, but certain difficulties arose on account of B[arbe's] death, which urgently required my
presence. I hope I can complete those transactions without losing much
time, so that I'll have at least September available. In Stockholm, I was
ailing all the time, either on account of the journey or because of conta-
gion, I can't judge, but I was down for days – those are the harbingers
of what is to come, dear Sofferl, and so I would like to enjoy some peace
and cultivate my inner spirit rather than wretched business matters. But
no, fate is stronger than our will, and so it goes on and on until business
comes to a stop as in the case of B[arbe]. We build forever on nothing
but quicksand, and the older we get, the looser the ground becomes.

True, you have caused me certain worries and were never able to
understand properly that I have acted towards you with kindness and
generosity, and yet I suffer when something is wrong with you and
when you are feeling so lonely in your life. Unfortunately you never
learned to spare my sense of honour, a point in which I am very sensitive, and that caused some misfortune for you too. But the past is past
and will never be recovered. Make sure to care for your health now and
don't worry, because worry is poison for your little stomach.

I am a stupid fellow who always thinks of others instead of himself, even of people who have caused me all sorts of hurt and sorrow.
Sensitivity is my worst fault, but few members of your family, it appears, suffer from it.

I hope the weather in Ischl isn't too nasty, so that your stay there will
benefit you. The most important thing is to keep your equanimity, to
put away your hot temper, and not to expose yourself to a cold. Good
Dame Nature will look after the rest, with or without Carlsbad. Many
heartfelt greetings.

You are surrounded by so many physicians that it is quite impossible to
get well. I almost believe that your only sickness is running to the doctor.

172b

[1890][237]

Dear Sofferl,

I suppose you have already received my registered letter to your father.

---

237 Unnumbered and tentatively dated 1890 in Nobels Arkiv ÖI-5.

They are starting to go after me here because of my success in Italy, and there is no doubt that I will shortly have to move my laboratory to another country. Strange news, is it not? People told me long ago, but the thing appeared rather implausible to me. But now it did happen, and the strangest threats have reached me.

Heartfelt greeting,

Your old devoted
Alfred

I am not even sure of my remaining at liberty, but don't spread the news, not even to your own people.

173

Aix-les-Bains, 29 September [1890]

Dear Sofferl,

I am still stuck here and taking a water cure, but my health and my spirit are gone forever. I will send you what you requested, but you won't be able to count on me much longer.

I don't know where I could find peace, perhaps in America.

With heartfelt greetings,

Alfred

174

October [1890]

I enclose once more some blue papers. I myself am rooted here with endless affairs and don't know how to help myself. Since Barbe's death things are worse than ever. And yet I never needed quiet more than today. But perhaps it is possible to go [to Vienna] soon and I don't want to hear of business for a whole month.

The water cure in Aix didn't help me much. There is no cure against old age and fatigue.

Heartfelt greetings,

Alfred

Why do you make me older than I am? My birthday hasn't come yet.

175

21 October [1890]

Dear Sofferl,

Today is my birthday, but not a happy one. I am so besieged by business people that I don't have even a moment to myself and have no rest, neither at breakfast, nor at lunch. Things have taken a turn for the worse, and since Barbe's death business has become a completely mad affair. I keep waiting for the end, but in vain, until one day I will make the crucial decision and disappear without a trace and without leaving an address.

How is your health? About mine I can't say more than that I don't have any. I would have liked a portion of health for my birthday, but no one granted it to me, and unfortunately one can't buy this noble and to me forever unknown merchandise.

With heartfelt greetings,

Your old Alfred

I hope it won't take much longer until I go to Vienna again.

176

29 October [1890]

Dear Sofie

I am lying in bed sick and suffering from an acute case of rheumatism. I can't write. I am sending you that certain something, but must insist that you put an end at last to your unreasonable waste. You are all crazy if you think I will continue like this in the long run.

Sincere greetings,

Alfred

177

[Autumn 1890]

Dear Sofie,

It's a fact. I can't believe that you and your family – especially your family – feel anything in their heart for me other than an interest in money. Or else you would have acted differently. No one can be in my company for five minutes without realizing how sensitive I am in matters of honour. And yet you tormented me for so many years by depicting me as a man who is not married but lends his name here and everywhere and sins against the most sacred human laws. And why? Because you know that I am too sensitive to counter such sacrilege with forceful means. Understand me well: I make no difference between a married and an unmarried woman if there is a mutual agreement and both parties are willing participants. But this is not the case here at all. I never wanted you as a lover and I have never permitted you to use my name. On the contrary I advised you to return to your parents and I absolutely forbade you to use my name. And what did you do? You gallivant around with lovers while introducing yourself as my wife. You can't do anything meaner to an honest man from whom you have benefited. If my name was less well known the matter would not be so important, but in this case I who am more innocent than anyone else have become the target of jeers and ridicule.

And now you want to tell me that you are interested in my health for my sake.

Believe me, you can stop your useless hypocrisy now that I have been told of that whole pretty mess you are in.[238]

And now I want to explain to you why I was so patient with you for such a long time. In spite of your frivolity and thoughtlessness there was in the beginning something fine and fresh in your nature. And at that time you were really fond of me. And then you could have turned into a quite good and decent wife. But you fell into the wrong hands and things went very, very wrong. Since then you have acted against me with a lack of scruple that is unparalleled and you still want to sing me a song of great devotion and gratitude. If you could be grateful

---

238 Presumably a reference to her financial difficulties in paying for her "palatial quarters" (see Letter 182). Sofie also became pregnant at this time, but she would not have been aware of her pregnancy before November. Nobel's letter, which refers to "winter coming," was probably from October or November, i.e., too early for rumours to reach him. He first mentions Sofie's pregnancy in Letter 188.

you would not be a true daughter of Israel, because on that side I have found only selfishness, over and over again.

I say all that only to explain why I have little faith in any interest in me on your part. One does not sully a man to whom one is attached. Rather tell me straight out that you've become afraid about the loss of money, for you and others, with or without title.

There is much filth in this world, but the winter is drawing near and white snow makes everything look innocent and beautiful.

Today, after a long illness, I feel a little better. It was an inflammation of the nerves in my head and neck. For many nights I could not sleep, not even with morphine.

Many sincere greetings,

Alfred

178

November [1890]

Dear Sofie,

When I sent you the two bits of paper, I was so sick that I could only write a few words with difficulties. Instead of thanking me, you write in an impolite fashion that I don't find appropriate at all. I am still in a miserable condition and sleepless and must not excite myself or get angry.

What you write about Olga agrees with what I heard about her.[239] I did my best to rescue her, not so much for her sake but because of your stupidity in introducing her to me as your relative. As soon as I had those unfavourable reports, I naturally withdrew all support and now she has to get on with her life as best she can.

As far as I am concerned I am more and more disgusted with people who abuse my kindness and hurt my reputation. It takes more than ordinary brutality to want to harm a person who has sacrificed himself for others and never thinks of himself.

My recovery will be very slow, and I myself don't believe that I will fully recover.

With sincere greetings,

Alfred

---

239 See Letters 172 and 181.

178a

13 November [1890]

Dear Sofie,

That's all very well for Bonato[240] to say, but he clearly does not under-stand my ailment. It may not be as bad as it was a few years ago, but it is more persistent and will not give way in spite of what I do for it, or rather against it. It is my left lung again that gives me such trouble. Bonato consoles me by saying that all people who have long nails like me are fated to suffer tuberculosis, but I am only half convinced of this so-called fact. In spite of everything my cough gets worse and run down as I am I can't get away from here. The reasons are diverse obli-gations that I took on and that are a heavy burden to me now. When a man is working with innovative projects nothing is more difficult than predicting when he will finish them. Every day looks as if it would yield the desired result, and it is hard to say why weeks and months pass by until everything works as it should. My profession is difficult and strenuous and robs me of my peace of mind. How I live, how I eat, how I sleep – or rather do not sleep! My domestics appear to feel sorry for me and do not envy me. Indeed a person would have to be stupid not to understand that one may lead a much happier life than I do.

I was not able to answer your father's letter before today. Apart from my wretched health, which makes it impossible for me to work, I am pestered on all sides. As a result I am so nervous that I can barely hold the pen.

I still don't have a receipt for the last 1,000 that I sent you. In any case I have had no message from you in several days and begin to be uneasy. Even if you have gravely sinned against me, I still pity you when I know that you are not well. So please let me know and accept my heartfelt greetings,

Your old friend Alfred

---

240 For Bonato see note 147.

179

17 November [1890]

Dear Sofie,

Mr Gschwandtner[241] writes that the villa[242] is apparently not insured. You, in turn, have answered my questions repeatedly, assuring me that there was an insurance against fire when the villa was bought and that the premium was paid. Please answer me at once. What is the state of affairs?

G also writes that far-reaching repairs have been ordered, and after I sent him 600 florins, he wrote and asked for more money. I don't want to hear anything about building or repairs that I have not approved and ask you to send me a full report. In any case I don't want to hear about the Ischl villa and will sell it at any price. That cursed villa is much to blame for the ugly gossip, and I must remove my name as owner. It is high time to open the eyes of the world to the unjust blackening of my name. I can forgive everything but that. The older and weaker I get, the more I feel and disdain injury to my honour.

With sincere greetings,

Alfred

I request an answer concerning the insurance and the repairs ordered without asking me. My health is very poor.

180

25 November [1890]

Dear Sofie,

After I already sent 600 florins, Gschwandtner writes that he has paid a great deal for installing stoves and the like and demands to be reimbursed. He writes that this work was certainly not ordered by him but that the invoices were sent to him.

I am quite displeased with this and I also want to get rid of the villa for other, more important reasons. The gossip involving me has always

---

241 For Gschwandtner see Letter 68.
242 I.e., in Ischl.

been disgusting and is turning more disgusting every day. I don't see why a man who lives like a hermit and does nothing but help others should be made out to be a disorderly and degenerate man and is talked about everywhere. Indeed it makes me indignant and apt to completely put me off benevolence.

Whether you live an orderly life or not won't change the matter. It is cursed now, and all the virtues of the world together with every bit of soap can't wash it clean. The only lesson one can learn from this is that one mustn't do any favour to a woman, or one will only get ingratitude and insult in return. If I don't seriously punish such behaviour, I must confess that it is no favourable testimony to my character.

Given my seriously disordered health it would be very pleasant to be in Vienna, where I would enjoy more peace at any rate and better care, but you have made it impossible for me to stay there, and I can hardly understand that I was so stupid as to go there. To be married is good, not to be married is also good, but that in-between story you and your family have put together in Vienna can't suit any decent man, least of all a man with such a touchy sense of honour and morality as I have. But you don't seem to understand that. You couldn't be more insensitive.

Sincere greetings,

Alfred

181

30 November [1890]

Dear Sofie,

Tomorrow I travel to Brussels where there is to be a great lawsuit,[243] but I will remain only three or four days. In the meantime you can address your letters to Paris, therefore. Day before yesterday I got a surprise letter from Mrs Böttger from Berlin, in which she asks me whether I supported her daughter. I replied of course that this was no longer the case. She wrote that Olga's lifestyle seemed very dubious to her and that is why she went to Berlin. Really she is to be pitied, for I regard the

---

243 See Letter 169.

woman as honest although I don't like her character. It appears she has heard a very unfavourable report about Olga's current life and wants to save her from complete ruin.[244] Not an easy task, I think.

This whole affair is very unpleasant for me, not because of the matter itself, which doesn't really concern me, but because it revives the gossip in Ischl, Merano, and other places. I am determined to get rid of the villa in Ischl and ask you to be prepared for it. Certain people, to whom I have been kind, have blackened my honest name and character sufficiently, and now I want to be protected against further smears.

A lady who uses my name without my permission and runs around with diverse ne'er-do-wells, who presents a rotten child as my niece, and thinks I will feed all of Israel, furnishes rooms for young gentlemen, etc. – you should understand at last that this can't please me in the long run. You were once a very dear and good child, and I have never forgotten it, but you seem to have lost your understanding of what is right and what is unjust, and if one is surrounded by many leeches, one can't be surprised if matters come to this pass. Fundamentally, you are still much better than the people in your surroundings and I am therefore very sorry for you. But all of you must have lost your sanity if you thought it could go on like this, and that the whole Jewish tribe in Vienna could live at my expense. If you want me to go on looking after you, you must first and foremost use your own name. It is stupid that you ever acted differently and it has harmed you in every aspect. I would have shown much more consideration otherwise, but I cannot allow an attack on my honour to go unpunished. Every decent person thinks about that as I do. But you are such a little helpless creature that I still pitied you against my better judgment and therefore imposed on myself true torture.

Sincere greetings,

Your always devoted Alfred

---

244 See Letter 178.

182

[December 1890]

Dear Sofie,

Sin carries its own punishment, and that is quite in order. You have furnished your almost palatial quarters without my approval and now you think you can maintain that mad luxury at my cost. Great mistake, as you will discover. Although you don't have Jewish traits in other respects, you are touched by a need to brag that is characteristic of your people, and therefore you long for a cage that is quite unsuited for such a little bird. But really it wasn't you yourself but a young gentleman who helped you decide, and that circumstance, which was once again couched in lies and deceit, doesn't make me any more forgiving.

I would have liked to come to Vienna, and my semi-home there was rather to my liking, but you and your family have made it impossible for me. You have, perhaps not deliberately, ruined my reputation so completely that it will hurt even unto my grave. A pretty reward for a person of my pure character and my morality.

I cannot hate, but I would rather not say what I think about people who have acted in this manner. In your case it was stupidity and lack of reason rather than malice as it was in the case of the others. But let us keep silent about it and let us hope that the punishment will not be as harsh as you deserve.

To judge by a short letter from Mrs Böttger, it's all over for Olga. She has sunk to the bottom, I think. But this is not very surprising when you think how she behaved in Ischl.

Sincere greetings, dear Sofie, from your devoted Alfred

How is your ne'er-do-well?[245]

---

245 Likely a reference to Nikolaus Kapy von Kapivar, who fathered Sofie's child, born 14 July 1891. See Letter 188.

183

13 December [1890/1889]

Dear Sofie,

I don't want to spoil your Christmas celebration with reproaches, but I need to make you understand one way or another that there has to be an end to the disgusting gossip. Every day, indeed almost hourly, people torment me with comments and hints that poison my life. They smiled significantly, adding a few comments about the apartment he furnished for himself. That is very pleasant for me, isn't it, and enhances your benefactor's life.

Then there is the miserable story of the use of my name, and as long as that goes round, it can only become worse. Can't you muster a bit of courage and move away from Vienna? You cannot lead a very happy life there, can you? Live in [Buda]pest or wherever you want, but don't force me to use means that go against my tenderness.

This name-fudging harms you more than me, and that is saying a great deal. After all you will lose your last support that way.

It is possible that you live in an orderly fashion now, but unfortunately that does not change matters. The harm has been done. There is no soap and no laundry to clean a reputation that has been sullied. Even your most innocent action appears tainted to the point of being talked about, and you have no tact.

I would have liked to go to Vienna at Christmas. I like it and find it homely. Here I can hardly remember the address of the theatres any more, whereas in Vienna I enjoy going to the Burg.[246] But you have made a stay in Vienna impossible for me and apparently don't even understand why – that is the height of stupidity.

And now I wish you with all my heart a merry Christmas and hope you will not find yourself lonely as I do. Many friendly greetings,

Your devoted Alfred

---

246 The Burgtheatre, still the foremost theatre in Vienna.

184

Dear Sofie,

This stupidity has gone too far – always to be obliged to feel for others, even for those who embitter my life! I therefore sincerely regret not to be able to spend Christmas with you. Essentially it is you who must be pitied, because you have a little heart that is good and your caprices might not have been so bad in other circumstances, but I repeat – you have made it impossible for me to stay in Vienna. I suffer too much from the gossip and am constantly reminded of the nasty reward I got for my kindness and forbearance such as can hardly be found anywhere else.

You must change in the coming year, dear Sofie, and give up your unreasonable desire to brag. It is not so much your own nature that is at fault or your very expensive living quarters that cause you only problems and worries – I suppose it is the work of those ne'er-do-wells rather than your own. If you weren't better in essence than those "gentlemen" I would have long stopped helping you. But in 1891 you must forget about the idea that my support will continue to the same extent as so far, and that that crowd of idlers will be permitted to live on my money and amuse themselves. On account of your weakness and helplessness I will help you once more, but not if you stay in Vienna where you are being fleeced on all sides. Open your little eyes as long as there is time to do it and consider earnestly how mad it is to think I will continue to look after the Jewish community at large and in addition after your ne'er-do-wells.[247] That must stop, my dear child, and very soon. Close your door to those ne'er-do-wells, put away your assumed name,[248] obtain living quarters that are not aimed at being palatial, that is, awful, and explain to your family that they must work – and then pleasant days will once again light up your life.

As it is now, your youth passes in an idiotic and stupid manner, without friends, without love, and you might very well find what you are looking for, if you weren't looking among worthless people who fleece you.

And now, dear Sofie, let me forgive and forget much (though unfortunately not everything) in your past. I wish you from my heart joyful

---

247 I.e., Sofie's family. Her father admitted that she supported him for many years. See Appendix, p. 275.
248 I.e., calling herself "Mrs Nobel."

holidays and a cozy Christmas celebration among those of your family who are truly good. Bertha is one woman whom you misunderstand – she is perhaps your best friend in Vienna, for she is cultured and has suffered. Such people stand head and shoulders above the crowd.

Best,

Your Alfred

185

22 December [1890]

Dear Sofie,

I am sorry I can't spend Christmas with you. I can imagine that you would have liked that and feeling for other people is truly my passion. But diverse reasons keep me from making the journey. I have a serious cold and could not very well embark on a journey before I feel better, and especially not on such a long journey. But the most important point is that you people have made my stay in Vienna impossible. In any case I am so entangled in business here that I don't even know when I can get away. We'll make arrangements, however, to meet at New Year's or a little later, but not in Vienna, because as I said, that place is impossible. For some time now I have had enemies there and they would pester me at the least opportunity. If I went to Vienna, I would have to stay at a hotel, and that would not serve your purpose. I have often done out of kindness what went quite against the grain, and will have to suffer my whole life on account of my natural gentleness.

Yet I am sorry that I can't see you at Christmas. I am keeping for you a small Christmas present, which I want to bring you later in person.

And now I wish you from my heart a very joyful holiday and no re-grets for the past injustice you have committed against me. I am living all by myself now and associate with no one. In fact I should visit a sick relative in Bordeaux, or rather Arcachon,[249] but I'm not in the mood for it.

All the best,

Your old Alfred

---

249 A spa near Bordeaux.

186

Saturday 27 December [1890]

Dear Sofie,

I spent only one day with my sick nephew in Arcachon and came back here yesterday, where I received your little letter.

Although I have been gravely hurt by you and your family, I don't want you to be hurt in turn if it can be avoided. It can't be avoided, however, if you stay in Vienna. The crazy things you did in the past will come back every day to haunt you. And why stay in Vienna? Really you have no friends there and only worries and problems on account of the life you are leading. You see things that torment and plague you without being able to change them. There are so many cities where you could be content and live a respectable life and, if you wanted to see the imperial city from time to time, it would be possible. There is Berlin, Dresden, Munich, Frankfurt, Brussels. Brussels is a charming city, where everything is available and one can live very comfortably. [Buda]pest too is very pleasant. Certainly you will see a lot more of me in a place other than Vienna, where stupid gossip about the past torments me and puts me into a bad mood.

Note that every time you did not follow my advice, things went badly. And it will be like that in future as well. You should stay where you don't have so many relatives. You should have a housekeeper and adopt a small child. You should also have a second dog because Bella needs to be kept young. You should live where you can pass as a young widow and you would soon find that life has its bright side. You must arrange it thus that people can't gossip as much, and that stupid change of name[250] has prompted a great deal of gossip and harmed me excessively without bringing you anything but unmerited disrespect.

You must notice that only one person means well and acts out of pure pity for your weakness and good heart. All the others, dear child, fleece you more or less.

My cough worsened on the journey and I will have to keep to my room for a few days. In the meantime the terrible cold will hopefully abate. Even if I can't wish you a happy and joyful New Year on New

---

250　To "Mrs Nobel."

Year's Eve, I hope to do so in the first days of the new year, and will do it also in advance, in writing only but from the heart.

I wish you good cheer,

Your devoted Alfred

187

Saturday [1890]

Dear Sofie,

As far as I can judge it seems that you are forever getting into deeper trouble and that everything will have to end badly. I see only one thing that can save you: if you had someone with you who has the necessary understanding that you totally lack. In that respect I know no better way than to be reconciled with Olga,[251] who seems to be made for you. She has judgment and you have feeling, and you two are made for each other. But you must set aside your caprices because no one, not even a domestic, can stand that.

I am completely serious when I write this. If you don't come to your senses soon, things will end badly. You can't be so stupid as to persuade yourself that I will go on like this, allowing myself to be fleeced. And then what will you do? You will be completely ruined, and you can only tell yourself that you thoroughly deserved that fate.

Well then, what do you say to my proposal? Olga would be a much better support than your family, and she would be fonder of you, because she has certain good qualities and is less egoistical than your relatives. Her address in Berlin is Wichmannstrasse 21.

Today I sent you a registered letter.

Heartfelt greetings,

Your devoted Alfred

---

251 This suggestion casts doubt on the dating of the letter. It is unlikely that Nobel would advise reconciliation after criticizing Olga's lifestyle in Letter 181.

187a

December [1890][252]

Dear Sofie,

I have every reason to believe that you are keeping from me a circumstance or rather circumstances,[253] and that can only harm yourself a great deal. In Vienna, especially in the circle of your own acquaintances, there is much talk about it, and if there is any truth in it you should understand how necessary it is to be candid with me since I will find it out sooner or later. This terrible secrecy is to your own disadvantage. I am sure you understand my meaning and will no longer hesitate to serve me up the true story.

Although I am breaking my back here, I can't manage the work before me, and I am in the grip of an incredibly nervous state. No one in this city of 2 ½ million people lives the way I do and torments himself with work as I do. Now I have such a longing for peace that I won't be able to stand it here for many more days. I don't know where I'll go, but far, far, away from the cursed noise of this metropolis, to some place where I can live like a hermit and far from the torments of hell to which inventors are exposed every day.

Many heartfelt greetings,

Alfred

Many thanks for the little folder [for calling cards].[254] But I don't visit anyone and don't need a folder.

---

252 Tentatively dated 1892 in Nobels Arkiv ÖI-5, but Nobel seems to be alluding to Sofie's pregnancy (see note 253). However, the letter may not have been sent off until January since the postscript answers Sofie's question whether he had received a folder for business cards (SH 2, 18 January 1891).

253 *Umstände*, seemingly alluding to Sofie's pregnancy since the German for "pregnant" is *in anderen Umständen*.

254 See SH 2.

188

[December 1890]

Dear Sofie,

The letter I have just received is full of regret and in a tone that is very different from your earlier half-insolent letters that you sent me from Ischl and elsewhere. But the main point is missing: a direct reply to my question, which I believe was clear. In Vienna and even in the circle of your close acquaintances they say that you are pregnant,[255] and since I must dot my i's and cross my t's, I will be completely clear and ask you again to consider the negative consequences that might and indeed will arise from your silly cover-up. In my eyes there is no greater sin than lies, and a cover-up would be the worst lie in this case.

You want me to forgive and forget, but you don't seem to understand what it is that I can't forgive and forget. It is human and understandable that a weak frivolous creature like you, left to herself and exposed to temptations, will go wrong. You and that gentleman, not to say, you and those gentlemen, behaved to me with a lack of consideration and disrespect for which one can't find mitigating circumstances and of which only people of very base thoughts are capable. That is what is so harmful for you, and your own conscience, which is perhaps awakening now, will tell you one day what great injustice you committed against your old Alfred, who nevertheless sends you greetings from the heart.

189

[1890][256]

Dear Sofie,

I want to know whether you have any personal belongings in the villa in Ischl, which should not go to the buyer, in which case they will be removed.

---

255 In January Sofie begged her sister Amalie to write to Nobel on her behalf and ask his forgiveness for her infidelity and the unfortunate consequences (see Appendix, p. 282).

256 So dated in Arkiv ÖI-5, but perhaps better placed in the year 1891 when Sofie's father inquired about the intended sale (HH 18 in Arkiv ÖI-5).

I am up to my ears in a lawsuit here and in difficulties of all kinds and don't know where all this will end. It is a regular robbers' den, where I will lose not only my possessions but also my life.

Best,

Your devoted Alfred

190

1 January [1891]

My heartfelt wishes [for the new year]. You are completely right when you say you are unhappy and lack a true purpose in life. But how can one go about things in such a wrong way, as you have always done? And now, judging by your terror, you are once again on the wrong track. One can only like a child if one likes the father or at least does not disrespect him.[257] But you want to take life by storm and you won't succeed. One must obtain a man's affection gradually, and then life will shape up as one would like it. It is strange that you yourself are very thoughtless and dependent and yet hardly ever accept the advice of another. That is the source of all your misfortunes in life, for to speak candidly, few people have had such opportunities to be happy as you. With a little understanding for what you owe to others and a little consideration for their honour, you could have had the best life, a life that could not have been better. But now it has come to this, and why? The moral to draw from all of this is that you must reward kindness and gentleness and that you yourself will benefit most from such behaviour.

Make an effort in the new year and completely leave aside your spitefulness, and then things will turn around completely. Truly, many people consider you worse than you are, and that is the result of your completely thoughtless actions.

My health is still very poor, and the after-effects of the illness seem worse than the illness itself. By the way, all the world here is sick, and the mortality rate is twice what it was in the time of the cholera epidemic.[258] The old year has ended badly and the new year is making a nasty beginning.

---

257 Presumably Sofie has owned up to her pregnancy.
258 See Letter 80.

Many sincere greetings and wishes for your well-being in the new year. I am very forgiving towards all people and very forbearing and don't hold any grudges other than concerning the harm done to my honour.

Again, all the best,

Alfred

Concerning your suitcases, I will arrange everything as soon as possible. But won't you have difficulties with customs there? Tell me about that before I send them off.

---

191 is now Letter 161a

192

10 February [1891]

Dear Sofie,

When I read your last letter, a word was on my lips that I will graciously spare you.

I don't accuse you of what you assume but of certain other things that I need not repeat. If someone abuses a man's kindness in this way, life is not worth living. I have often and repeatedly said to you and your family that this will come to a bad end, and all of you have begged me for years now to be tougher [on you]. Such lack of consideration – no, that is too delicate a word – cannot be found even in novels.

I want to know: what else have you committed to sully my honest name? It must be something awful since my acquaintances have not returned my New Year's wishes.[259] That's the result of my goodness. I will remember it.

Your devoted Alfred

I enclose some bits of paper – Italian.

---

259 See SH 4.

193

[First half of 1891]

Poor little child,

You need a few comforting words rather than reproaches for past deeds. Your missteps I attribute especially to your education and your surroundings as a child. You have a very small soul, but not a bad soul, and I would not have recognized your Jewish patrimony, had it not been for your complete lack of consideration. Don't be disturbed and don't worry because that is bad in such circumstances.

Is your sister Amalie[260] with you? She is good and sympathetic. Perhaps not quite as mild in her judgment as I am, but that isn't surprising, since my mildness is unique.

From the heart,

Your devoted Alfred

194

18 May [1891]

Dear Sofie,

A serious cough is not a "jute Jabe Jottes" as they say in Berlin.[261] The sooner one is rid of it the better.

But I won't go to Ischl, for several reasons. You must be a little dim-witted not to see the reasons yourself. You could not move there without exposing me to hurtful remarks. And you can imagine that this business does not suit me at all. In any case I expect to sell the villa in the near future.

You ask where you could go? Switzerland isn't a bad idea, but not where there is a crowd of people. You need quiet and you can find that only in small places, for example on the shore of Lake Constance, in Lindau, Romanshorn, or similar places might be very good, I think. I

---

260 See Amalie's letters to Nobel, Appendix, pp. 282–4.
261 Berlin dialect for a *gute Gabe Gottes* – a gift of God.

will consider the matter and write to you soon. If you go to Switzerland, I may meet you there at some future time.

I enclose a cheque that will make the matter easier for you. Please don't abuse it.

Best wishes,

Alfred

195

[July 1891]

Dear Sofie,

From your letters it seems that you have suffered gravely,[262] poor creature, and so you are seriously punished for some of your earlier sins. I am very forbearing and if you hadn't done anything to damage my honour, you would certainly have found a very mild judge in me. But there is one thing I can't forgive – when someone tries to make me ridiculous. The resulting bitterness does not decrease with the years but rather grows. I hope you will finally learn a lesson from the past: if one lives off the good deeds of a decent man, one mustn't defile him. If your family knew how you carried on they would certainly be even more surprised that I am still helping you in some form.

But now that you have suffered so much, you will perhaps understand the noble thoughts I am capable of and the reward I received for them.

Many heartfelt greetings,

Alfred

---

262 Presumably a reference to her difficult delivery on 14 July. See SH 8 and note 22.

196

17 July [1891]

Dear Sofie,

It would not be impossible to come for a few days to Gmunden,[263] but first I have to put up with a long and tiring business trip. These trips are certainly no fun for me, for even the shortest journey affects me like a disease. The clockwork of my body has been greatly disturbed.

You keep writing about my gentle feelings. Gone are the times when you told me that I am "a little brutal."[264] I wonder in what this "brutality" consisted.

You are certainly not brutal by nature, for your three sisters have inherited something refined and delicate from your mother. How that poor woman must have suffered in those surroundings – a gem in a setting of filth!

Heartfelt greetings,

Alfred

197

Lindau, 11 August [1891]

Dear Sofie,

I have preached so often, asking you to be reasonable, but without success. It would therefore probably be a waste of words if I repeated my well-meant advice.

If you continue with your wasteful life as before, you will soon be beggared, and that is not a pleasant prospect. Your brother-in-law lives with his wife[265] and two children quite well on less than 2,000 florins a

---

263 If the year is correct, it is surprising that Sofie would consider travelling so soon after the birth of her child. Gmunden is about 250 kilometres west of Vienna. Sjöman (*Mitt hjärtebarn*, p. 303) notes that Nobel registered at a hotel in Bad Kissingen on 22 July 1891. The registration read "Mr Alfred Nobel with son and daughter, from Paris."
264 See Letter 121.
265 Albert and Amalie Brunner; see Letter 134. For Amalie's letters to Nobel see Appendix, pp. 282–4.

year. Rothschild may accept your silly statement that you must have 30 times more, but it doesn't work with me. In my eyes that is nothing but an importunate way to abuse my goodness. A long time ago I alerted not only you but your father as well to this mad situation, but it seems that all of you have lost your heads.

Now, dear Sofie, your craziness has resulted in another human creature depending on you,[266] and I assume that this will enlarge your microscopic brain a little. Therefore prepare yourself in time to live in modest circumstances, like your brother-in-law, and take care the many leeches don't suck you dry. You can count on my help only under those circumstances, or you will fare badly, very badly.

This time, don't waste my well-meaning advice that arises from warm feelings of sympathy.

Heartfelt greetings,

Your devoted Alfred

198

Genua, 2 February [1892]

Dear Sofie,

First of all you must send to Paris a clear and detailed account of your debts.[267] Then we'll see whether I decide to do something to help you out once more and get you out of your difficulties, which are the result of your stupidity. It sounds incredible that you managed to waste so much money and in addition get into debt. How those low-lifes must have exploited you, and how stupid you yourself must have been, even more stupid than I believed. Really, you do whatever is possible and even what is impossible to bring misery on yourself. In former years you weren't as silly and childish. Your brain seems to become smaller. How do you think your future will shape up when you carry on with such madness??? Do you believe that I will, in my kindness and generosity, allow a stupid woman to ruin me?

---

266 See SH 8 and note.
267 See SH 11, 17.

I have warned you long ago, but it seems you must experience poverty.

Sincere greetings,

Alfred

199

24 February [1892]

Dear Sofie,

You don't have the moral courage to keep away from stupid luxury and stupid people who make your life bitter. Why didn't you exchange your large quarters for smaller ones long ago? The mad desire to boast will ruin you. You have, as you say, a little daughter who gives you joy. I hope she will give you a purpose in life and comfort as well. You want me to forgive you. That is impossible. I cannot erase years of moral outrage I felt at the sins you committed against me. There are roots to which I am attached forever.

It was not to harm you that I gave so much information to Dr. Barber.[268] On the contrary, he can give you good advice only if he knows the entire situation and my true intentions. I have written to him today the conditions under which I will pay. In fact it is a stupid thing to do, because in my opinion you are beyond rescue.

Did I not tell you long ago that I suffered huge financial losses last year? That should have opened your eyes and put an end to certain carryings-on. But it cannot become apparent to a person who has no conscience. You must cut your ties to the circle in the Praterstrasse[269] and associate with someone who has intelligence and discipline. Then perhaps your better qualities would come to the surface.

Wishing you all the best for your well-being,

Alfred

---

268 Answering Sofie's complaint in SH 10. Maximilian Barber was her lawyer in Vienna.
269 Heinrich Hess lived at 50 Praterstrasse.

200

16 March [1892]

Dear Sofie,

I enclose 2 Hungarian bills, valued at approximately 220 florins, and in addition 1,000 francs, to cover what you owe to Mr Trensch,[270] as you wrote. You have fallen into a pretty hole and have nice relatives who helped you in that process. All of them apparently lack any moral or legal understanding.

Now go at once and pay in the presence of a witness and ask for a receipt of the draft and send the receipt or receipts to me as soon as possible by registered mail. You should do all these transactions directly without involving Mr Barber. Make sure everything is done as I tell you here.

Your devoted Alfred

201

29 April [1892?]

Dear Sofie,

I enclose once again 2 Hungarians, assuming that you have come to like this nation.[271] The arrangements I have to make here demand more time than I calculated, and I don't know exactly when I can depart from here. It is probably better to address letters to Paris, for all letters are immediately forwarded to me.

I hope some reasonable person – that is to say, someone outside your family – will make sure that you change your ways in time and don't become a beggar. I have warned you of this for years now, but as the proverb says: Even the gods fight stupidity in vain.

Sincere greetings,

Alfred

---

270 The debtor Sofie mentions in SH 11.
271 The father of Sofie's child, Nikolaus Kapy, was Hungarian.

202

Paris 18 May [1892]

Dear Sofie,

They say that Christ was a very good and forbearing fellow. But he had a dear Papa who was able to reward him properly for it and ever since he is said to live in heaven, "like God in France." But I have no such exalted relatives and can't see why I should be asked to be 10% more benevolent and forbearing than God's dear son. Furthermore, His Excellency surpassed me greatly in the table of multiplication. As soon as anyone teaches me how to make of three thin fish 2,000 fat fish in a trice, without hocus pocus, I will buy 3,000 fishes and will turn them into 2 million, which will make me a lot of money. But I will be more generous with it than the Crucified, and then God will perhaps be angry with me because I surpass the generosity of his First-born. In the meantime I plan to show you and your family that I am very kind but have no intention to be fleeced further. Please don't overlook this little word.

Your most devoted Alfred

---

203 is now Letter 178a.

204

[San Remo,] 15 December [1892]

Dear Sofie,

You complain. By contrast it seems to me that only very few people are as lucky as you. You have done everything possible to make me deeply miserable and if you had not met such a kind and benevolent man as I am, what would have become of you? You would have come to a miserable end and perished. So there is no reason whatsoever to lament.

My heart disease is for some time now very bad. I don't know when I'll reach the end, but the harbingers are unmistakable. The heart trouble is accompanied by a dull ache that greatly affects my already low spirits.

I wish you a merry Christmas with all my heart and as little blood-sucking from your relatives as possible. I don't know where I'll spend the holidays, but it's all the same to me, if only the heart problems would abate a little.

Sincere greetings,

Alfred

---

205 is now Letter 187a.

206

Nice 23 [December 1892]

Dear Sofie,

I add to these lines a wish for very cheerful and pleasant holidays, for it appears that you now have a purpose in life. If you take it really seriously, you won't find the time slow to pass. I am much bothered by my heart disease and am spending Christmas here in the utmost seclusion and solitude. I don't know a soul and don't even make an effort to check the list of foreigners to discover whether there are any familiar names.

Best greetings and wishes,

Your most devoted Alfred

207

San Remo [1892/3]

Dear Sofie,

From your last letter I see that you intend to keep your palatial quarters for another year. That is your business, and I don't like to interfere in the affairs of others, but I thought I outlined the situation very clearly in my last letter, so that it would not be difficult to draw your conclusions from it.

I am glad that you are better and send you best wishes,

Alfred

208

[1893]

It very much looks as if things in Vienna are once again going awry. It is really incredible that you refuse to understand that things can't go on like this. You should have given up your quarters long ago and made different arrangements. Really, you have no brain, and the little you used to have seems to have bit the dust as well.

Sincere greetings,

Alfred

Tuesday, San Remo

Write to my Paris address, as usual.

209

17 January [1893]

Dear Sofie,

I was too ill to write earlier. You were quite right to assume that all the great difficulties of the past year have affected me greatly. I have turned completely gray, not only on the outside but also on the inside.

Starting with the new year, the new order is in effect, and here is the first "branch." I hope what I was told recently is not true – that you have once again incurred debt. That would spell ruin this time. Therefore I give you a timely warning.

I received a nice letter from your sister in America.[272] Her children work and earn money even though they are very young. That is something very different from the despicable idleness in the Praterstrasse.[273] The sense of honour of all those people combined is less than Bella's on her own.

Your most devoted Alfred

---

272 Bertha, who lived in Chicago at that time. For the Hess family see Letter 132 note. See also Sofie's references to her sister in SH 14 and 27.

273 Where Heinrich Hess lived.

210

15 March [1893]

Dear Sofie

If I am to help you in some way you must first and foremost stop all deceit. In your letter a few days ago you mentioned 3,600–4,000. In the telegram I received just now it is 5,000.

On 2 August '92, you sent me a bill from Mr B[arber][274] in which he demands payment of 3,850, including his fee – as if that was my business!

Where does the 5,000 come from? You have been strictly forbidden by more than one person to incur more debt. Yet Mr B wrote to me already on 21 December 1891 – I quote – "I withdraw my services the moment I see that Madame can't make ends meet or incurs another 100 florins in debt."

Thus it is unthinkable that Mr B was involved in raising your debt. To require 5,000 florins since August '92 – less than 8 months – he would have had to charge you 45 percent/year.

Please tell me the truth. You won't get far with lies.

Your most devoted Alfred

211

[An alternative draft of Letter 210?]

Dear Sofie,

If I do any more for you out of kindness, it all depends mainly on you stopping all deceit. Take care therefore, for I know very well what to believe or not to believe.

On 21 December '91 Mr B wrote literally: "The moment I see that Madame can't make ends meet or incurs another 100 florins of debt, I withdraw my services." If Mr B lent you money after this written statement it serves him right if he loses his money.

In a letter dated 2 August 1892, you enclosed a bill of Mr B in which your debt to him, including his fee, is given as 3,850 florins. I don't have

---

274 See Letter 199.

anything to do with this affair and find it amusing that someone sends me bills that are not in my name. But since the matter has come to my attention in this way I know that the amount is 3,850 florins and not 5,000 florins, as you indicate now.

In any case, how does it come about that lawyers in Vienna lend money to ladies? Are they bankers there as well? Here that is not the case, as far as I know.

In any case, Mr B cannot have lent you anything since he himself stated in writing that he would not do it. How then did your debt increase from 3,850 to 5,000 florins between August 1892 and March 1893? Please tell me the truth or I would have to assume, counter to my wishes, that you paid a horrendous rate of interest. That can't be tolerated.

Please answer me immediately, candidly, and sincerely. It's no use to quirk your lips. You must come out with it.

Indeed I would welcome a lawsuit between you and Dr. N.[275] I would have it monitored by a jurist who would see to it that my unselfish and noble attitude will be appreciated and the massive insults that I was subjected to would come to light.

Your most devoted Alfred

212

[1893]

Dear Sofie,

Do you think you can make me believe fairy tales? A lawyer isn't an innocent boy who does not know what his client owes him or doesn't owe him. In July 1892 Mr B[arber] gave you a bill written in his own hand, which completely contradicts your current claim. This story isn't kosher, therefore, and since God, the pope, and the jurists are infallible, there is only one conclusion to draw – that you have muddied the matter on purpose. In any case Mr B has no other claim on you than his advance and a fee of 1,000 florins, and I assume you paid your debts. I have no time to write back and forth a great deal about this.

Your most devoted Alfred

---

275 Unidentified

212a[276]

San Remo 22 September [1893]

Dear Sofie,

I have just arrived here, but I am in such poor health that I write these lines with difficulty. I stayed much longer in Aix than I had intended, because the bathing weakened me greatly this time, so that I was unable to leave. On the journey I suffered the most terrible pains from colic and arrived here in a very wretched condition.

Sincere greetings to you and Olga,

Your devoted Alfred

213

Hamburg 2 October [1893]

Dear Sofie,

My heart ailment gives me trouble. I can't write therefore but send you my best wishes and enclose what interests you more.

Your devoted Alfred

Send your letters to my address in Paris.

214

Savoy Hotel London, 17 February [1894]

Dear Sofie,

I send you an enclosure, but can't write. I am too busy and completely fatigued from the strenuous journey. A number of things have been going terribly wrong for some time now. But I suppose you are too stupid to understand that.

---

276 Translated into Swedish by Sjöman, *Mitt hjärtebarn*, p. 326, from an original in the possession of Olga Böhm. A copy of the original is now among Sjöman's papers in the National Library of Sweden.

Best wishes

Your most devoted Alfred

Use my Paris address as always.

215

Aix 12 September [1894][277]

Dear Sofie,

My sudden departure was caused by that terrible misfortune in France. I had no time to say adieu.

I found you in better health than ever and don't know why you complain. But you do have a problem and your surroundings are neither very good nor very pleasant. But on the whole you are not one of the most unlucky people although you did all sorts of things to make that happen.

Your character has improved, and your unparalleled capriciousness has lessened somewhat. The fact that I avoided taking you into society was not the consequence of your current behavior, of which I have no criticism, but of that awful abuse of my name in the past.

You had incredible luck – anybody must concede that who is familiar with the circumstances. Anyone else would have abandoned you and left you in the lurch, something you tried hard to bring about.

By the way, it is a coincidence that I could do what I did. I hung by a thin thread. It might have come out differently.

Your little child is quite nice.[278] You must give her the right kind of education. I don't know anything about your relationship to her father and therefore cannot judge which of you is right or isn't right. In any case I have nothing to do with that.

Heartfelt greetings,

Alfred

Write to my address in Paris, as usual

---

277 So dated in Arkiv ÖI-5, but more likely from an earlier date. The "terrible misfortune" might refer to the raiding of Nobel's laboratory by the French authorities in 1891.
278 When did Nobel meet Sofie last? In February 1892 (SH 11) Sofie complains that she sees him "so rarely."

216

7 March [1895][279]

Dear Sofie,

Is it true that your Captain is willing to marry you?[280] Such an action would perhaps be not only right but also sensible. Only put away that stupid pretentiousness of yours and that stupid moodiness. Still, on the whole, you are a woman of real feeling, and that is worth something too. I even think that you are not completely without conscience as long as you keep a distance of 200 miles from the Praterstrasse.[281]
  Heartfelt greetings,

Alfred

216a

Undated[282]

My dear child,

In the afternoon I received your dear letter, and am very glad to hear that you are finally going to be sensible, and pay the price [for your errors]. It is high time for you finally to begin a sensible, orderly life, to

---

279 So dated in Nobels Arkiv ÖI-5, but more plausibly dated 1894, answering Sofie's SH 17 and corresponding to a letter of 16 Oct 1894 he wrote to his lawyer, Julius Heidner: "Does the lady want to get married to give a name to the child she is said to have with Mr K, then she can have him cheap. A man who is on her own cultural level should be eager to make the arrangements" (quoted in Sjöman, *Mitt hjärtebarn*, p. 73).
280 I.e., Captain Nikolaus Kapy von Kapivar (1856–1913). They married in September 1895. See SH 30–3.
281 I.e., from her father, who lives there.
282 Supplied to Sjöman by Olga Böhm. See his Swedish translation in *Mitt hjärtebarn*, pp. 62–4. The original does not appear to be extant according to information received from Kristina Sandahl at the National Library of Sweden, where Sjöman's papers have been deposited. Olga Böhm died in 1995.
   The content of the letter is cryptic. A possible interpretation is to regard the conversation related here as an internal dialogue, showing Nobel's conflicted relationship with Sofie. This would explain the contradiction between the writer advising her to move closer, and the unnamed gentleman wishing to keep her at a distance.
   I would like to thank Asa Andree for assisting me with the translation of the letter.

put an end to pretentiousness and begin to provide a home for yourself in old age.

It is not without reason that I preach to you, my dear child. You may already have heard from Alfred about the conversation we carried on during the whole journey from S. to here. The journey was one of the most embarrassing in my life although, or precisely because, it produced the most amicable explanations. Throughout the trip the talk was only about you. I guessed right away that something was amiss when he began telling me your life story and examined his relationship with you, saying that you were not married, etc.

I had to sit there and listen with the world's stupidest look on my face. I could already anticipate the end of the song. He reproached me, criticizing my relationship with you, which was already on everyone's lips, that I was constantly with you, etc. I am still amazed at my presence of mind and the answer I gave him. As a gentleman I had no choice but to take his hand and say: If it makes you uncomfortable, I will not enter her house anymore.

This was, briefly, the subject of our conversation, which continued during the whole journey. He must have been very upset because he smoked incessantly. I too was understandably very upset. Finally he calmed down and let me off at my house, but declined to get out himself.

This, briefly, was my experience, which is all the sadder as I want to keep my promise and won't visit you again. My promise is too sacred to break.

I give you good advice: Move closer to him and move to Paris. It would be too much to tell you everything he said. But among other things he said that he was overjoyed you were so unwise as to move away from him and from Paris. He intends to curtail your special benefits and install you in Graz. There you will live under your own name until you are married.

Throughout that time he talked of you as "little Hess."

Now farewell, my dear child, everything is done now, and act so that you make at least some amends for your offences against him.

I denied firmly any relationship with you. I said I came to you so often only out of pity, because I was moved by your helplessness. Then I also said that there could not be the slightest trace of love because one can only love people who have nobility of soul, which you completely lack.

So once again I beg and beseech you: be wise. Your fate is in your hands now. Do not frivolously put your future at risk. In any case I do not intend to contribute to this matter.

As far as concerns me, I shall walk alone and drag on with my life.

And now farewell. In my heart I embrace you fervently once again,

Your old man

PART 2

Hess's Letters

The ordering of the letters in the Swedish National Archive, Nobels Arkiv ÖI-5, has been changed to reflect a more cogent time line. The numbering in ÖI-5 appears in parentheses after the number assigned to the letter in this edition. To distinguish Hess's letters from those of Nobel, the numbers are preceded by the designation "SH."

I have retained Hess's punctuation to give readers an impression of her breathless style. Her run-on sentences have been broken up only where this was necessary for comprehension. The numerous phrases she underlined are italicized in the text.

Collation of Letter Numbers

ÖI-5	This edition
1	17
2	2
3	1
4	4
5	5
6	6
7	7
8	8
9	16
10	12
11	3
12	9
13	10
14	11
15	13
16	14
17	15
18	19
19	18
20	21
21	22
22	20
23	23
24	25
25	27
26	28
27	29

Collation of Letter Numbers (*continued*)

28	30
29	31
30	32
31	33
32	34
33	35
34	36
35	39
36	37
37	38
38	40
39	24
40	26
41	41

SH 1 (3)

Undated[1]

My precious Alfred, if I may still call you so, for you have not departed from my heart in spite of the endless and great torments I must suffer. It is only now that I realize that the accusation against me is non-existent[2] and is merely a sworn plot, and that you *readily believed* the wrongdoing they attributed to me, to surrender me to misfortune once and for all. I never ever thought you would act in this manner against me, who loves and respects you so much, by God I don't deserve such abuse.

I expected no such defamation from the person who spoke of me in this manner, for to bring accusations against a helpless person would be more than sin. God only knows that I am blameless and I believe he will not desert me even in my present condition and awful position. I cannot describe to you how much I suffer, and it would also be useless, but when one suffers such torments and is as downcast as I am, it is easier to write it down. I have not a single human soul to whom I could entrust myself, I am alone and deserted as never before, don't know what to do and how to go on without my only love, Alfred. You have been everything to me in this world, you have been torn from me as well, and for nothing, without any reason. You know best how devoted I was to you, how I respected you, obeyed you, your wish was my command always. I lived such a quiet and upright life, for you alone, for you only, whom I loved and idolized, and now I have to suffer such shame for it. Ah my God, I can no longer live without you, I am so weak, you would not recognize me again. I am half-mad. What shall I do, I can't leave you, I love you genuinely, with my whole soul.

---

1 The letter may date from 1879, when Nobel accused Hess of an impropriety involving his nephew, Emmanuel. A date near the beginning of their relationship would make sense since Sofie expresses a desire to learn French. The scenario also fits 1887, the year in which Sofie's father defends her against accusations of impropriety, mentioning the affair of 1879 and stating that the allegations "are based on vengeful statements of Olga" (Appendix, p. 278). By that time, however, Sofie's remarks about wanting to learn French seem ludicrous. She had been studying French, but with little success (see Letter 9). As for Nobel's suspicions, see Letter 135 of October 1888, in which Nobel admits that he had Sofie investigated in 1885 (?) and found that rumours about her indiscretions had originated in Vienna and elsewhere. There were no allegations against her in Paris.

2 She spells *existiert* "egsistirt."

Allow me to stay here until I have learned the language so that I don't have to commit myself to working in public.[3] I will go into a household, in whatever capacity, and earn my bread, but you are wrong to think that I will discard you. That will never happen, as far as I am concerned. When I took that step,[4] it was not to go on a spree, but because I felt drawn to you, I loved you even then and so it comes to pass that I am miserable. I don't blame anyone, but let me tell you: Even if you consider me for I don't know what, I can give you my most sacred word that I come from a very good family, and am therefore even more unfortunate.

Even if you allow me to stay here, you have no obligation towards me. I will rent a small room and work – but the only thing that will make me happy is to live in the same place as you, my precious Alfred, who will always be everything in my life. I have tortured you a great deal, and now I have to pay the penalty for it and be so wretched. When I depart from you, I claim nothing but your respect, which you have always given me so far. There can be no talk of pensioning me off. I don't need such alms, for I did not act out of calculation, God knows. Your presents will be returned together with everything else, which only makes me wretched now. I never believed I'd become so unhappy. If there is a divine power, my innocence will come to light.

Ah my God, if I could only see you, I am beside myself, my whole being, my home is so empty and deserted without you, oh hear me, if you are not completely without feeling, and you are so good to *everybody*, you are so hard only against me, and without blame or reason.

What can I do to see you just for a moment, that is all I want, and to press my lips to your hand.

I want to tell you so much. Perhaps you will then believe in my innocence.

Hear me and come.

---

3 *Daß ich mich nicht mehr den öffentlichen Geschäften hingeben muss.*
4 I.e., to join Nobel in Paris.

SH 2 (2)

18 January 1891

Dear Alfred,

In the past three weeks you haven't written a single line to me, I don't know why, it's not like you not to answer and to leave me worried. I wrote three letters to you – to Paris, that is – I didn't know that you are already in San Remo and the furnishing of your villa is completed.[5]

In Vienna they talk about the splendour and elegance of your house. I wish you much happiness in your new home. I feel bad about the fact that you have never given me your new address and haven't even written to me about the villa, I had to find out from strangers.

How is your health, I hope the fine air there will do you good, and as I hear, you have company, the Viennese know everything, it is a nest of gossip, sweet to your face, false behind your back. I would like to ask you, dear Alfred, to send me some cash. For the past fourteen days I have been without money and had to pawn my last ring to live, that's what I have come to. I saw Olga[6] here three days ago at the bake shop, her brother Klaus ("Little") was with her and the fat actor, she has been here for a long time, dressed in mourning, has her mother died? She pretends not to know me, *the stupid thing*, probably has good *reasons* to hide out. As far as I am concerned, she can live with ten men, that's none of my business, she must do well, she looked very elegant. Apropos[7] – did you get my folder for business cards, I would have thought you'd mention it? I worry about food, I don't know where I should get the money for it tomorrow, I am ashamed before the servants, especially Toni, whom you know from Baden, it's a tragedy and wretched not to have even ten florins and be in this position.

Many greetings and kisses,

Your Sofie

---

5 It seems that Nobel rented the villa on Corso Felice Cavalotti, San Remo, at the time. He signed the contract of purchase only on 25 April 1891 (http://sanremoguide.it/en/the-city-of-san-remo/villas-and-gardens/villa-nobel/).

6 For Olga Böttger see Letter 33. Presumably she was on a visit in Vienna. She lived in Berlin (see Letter 187).

7 Here and elsewhere spelled "aprospos."

SH 3 (11)

Saturday 1 Feb 189[1][8]

My dear little boy,

I was more than stupid. Something like this can happen only to an unlucky person. I have quite *burned my hand* and can write only today, with an effort. I took a bottle of water and poured it into the stove to quench the fire. A great cloud of steam shot up and scalded my hand. I've never experienced such pain. It's terrible, let me tell you, dear Alfred. I had my hand immediately bandaged and even now can't use it. I didn't sleep a wink for two nights because of the pain being so terrible, and with all this I don't have a maid or a cook, a nice situation, everything unpleasant is added to my horrible existence! That's why I only get around to writing you today, dear Alfred, and thank you for your dear words. I am repentant and accept what we agreed on, for I realize very well that you are right and the fault is mine, perhaps I am too good natured or too *stupid*, either one, and I promise (I mean a holy promise) I'll not borrow one more penny, then I won't live such an uneasy life, and peace of mind is all I can hope for.

In any case I'm constantly in a tizzy, that's probably a consequence of my condition, and my dearest wish to have a child is spoiled because of all those accusations and insults, so that I'll give birth to a sick creature instead of a child, for all I know it will end up looking *like Babs*[9] because I don't have anyone else to look at. I take no joy in life any more, dear Alfred, and you being so hard to me, that makes me completely sick.

---

8 The year date is based on the mention of her pregnancy. The day of the week (Saturday) does not accord with the year date, however, and must be an error.

9 Dr Barber? See Letter 198. Alternatively, *Babs* may be a dialect word, meaning "muck" (according to the *Pfälzisches Wörterbuch*: http://woerterbuchnetz.de/PfWB/?sigle= PfWB&mode=Vernetzung&hitlist=&patternlist=&lemid=PB00041#XPB00041).

SH4 (4)

Friday 14 Feb 1891[10]

Dearly beloved Alfred,

The short letter I received from you yesterday[11] disturbed me very much although your reproaches are all quite justified it hurts me terribly that you are so against me, I don't deserve that and if you saw the way I live now you would judge me very differently and would go easier on me.

I bitterly reproach myself, dear little boy, and I repent everything you accuse me of, but too late, as you say yourself. Now you will realize, dear Alfred, that I deserve pity, for I have nobody who is willing to look after me at all, not a soul, let me tell you, only when they need money they look me up, all of them, except my sister Bertha,[12] not her, all the others have no heart, only when it comes to money, they don't care where it comes from, and I find that *nasty*.

Many thanks for the four Italians,[13] and today for the 4 Hungarian bills of exchange, I got 4,000 florins for them and give you a thousand thanks for them, dear Alfred.

I will pay the two seamstresses with the money, the others will have to wait, it's in any case only the women who are so nasty and so pressing. The other merchants aren't so pressing.

What you write about "sullying," I don't understand,[14] I can't help it if your acquaintances don't send you cards, that kind of thing has nothing to do with reputation, because there are people you don't want to touch with your toe, and yet they are great in this world, you are often too scrupulous and believe immediately the worst, you can rest assured, nothing is happening that could harm you, dear Alfred. And now let me ask you, how you are and what you are doing? Ah, if I weren't in this condition, I'd immediately leave here for I feel so lonely and unhappy as never before, that is going to be a nice child, what with all this trouble and sorrow all the time!

---

10 Once again the day does not accord with the year date: 14 February 1891 was a Saturday, not a Friday.
11 Nobel's Letter 192 of 10 February 1891.
12 For Bertha see Letter 147 and note.
13 See Nobel's Letter 192, in which he mentions sending her bills in Italian currency.
14 In Letter 192 Nobel asks what Sofie has done to "sully" his good name and complains that his acquaintances have not answered his greetings.

With a thousand kisses and greetings and once again many thanks,

Your Sofie

SH 5 (5)

Monday, 30 March 1891

My dear Alfred,

As of today I still don't have a reply to my two letters, I'm awaiting it with great longing because I don't know what to do, I will follow your advice and am only sorry I didn't follow it earlier, I would surely be in a different position and would have to suffer much less sorrow, for to expect a child in such circumstances is hard, that I can tell very well and wouldn't wish it on my worst enemy, for what I am suffering is terrible, you wouldn't believe it, dear Alfred, because not all men are as sensitive as you and as good as you, dear Alfred.

Have you already written to Olga? Anna[15] came back from Berlin yesterday and told my father that Olga looks very old and ravaged and has completely ruined herself. It is said that she lives with many men.

I am very sorry for her because in the end she will be the most unhappy of women and have no support.

Am I not right?

Write soon. Heartfelt greetings and kisses,

Your Sofie

SH6 (6)

Friday 10 April 1891

Dear Alfred,

I received your telegram from San Remo and was quite surprised, I thought you were still in Paris, your days in San Remo must be very

---

15 For Olga Böttger see SH2 and note. Anna was Sofie's half-sister, daughter of her father and his second wife, Julie Stern. She was admired and used as a model by painter Hugo von Habermann, but suffered from paranoia (see Sjöman, *Mitt hjärtebarn*, p. 89).

nice and likely warm, here we've had continual rain for the past fourteen days and I would be happy to get away from Vienna for once. What pleasant days and hours I had when I lived with you and travelled with you, yes, even the *gods fight* stupidity in vain, that proverb is correct. Unfortunately I came to my senses much too late and must pay for it now, well, who knows how long it's going to be, I'm always half-way to the grave, especially since I have no one. A mother is the main thing, and I don't have one, and there are circumstances that contribute to my unhappiness, being unmarried and having no other status. In short, all those things make me sad and my health suffers accordingly. You are the only person who is good and feels for me, but you too have changed greatly recently and don't care about me, I don't belong to anyone and am quite alone. How is your health, dear Alfred, do you have company or are you alone? Have you written to Olga? Anna told me recently that Olga is having an affair with a piano manufacturer there[16] and lives luxuriously, well, you know those *uncles*, but they say all her dresses and all the things she has come from you, that's what Hermine[17] heard from some actresses. I always thought Anna would visit her, I was curious to hear about her looks.

I paid Dr Schwarz[18] cash from the money you sent me.

---

7 (7)

Monday, [early summer] 1891[19]

Much beloved Alfred,

Many thanks for your lines and the content of 2,000 francs, which I received yesterday. I have been without money for some days now, had not a penny. There must be very few women in my position, without a penny, neither bills of exchange, nor savings, nor cash in hand. When I think about my future, I shudder, wondering what will become of me.

---

16 In Berlin.
17 Hermine, Sofie's stepsister, worked as an actress in Berlin, but had to content herself with small roles. See Letter 140.
18 Not identified.
19 The content of this letter is similar to that of SH 5 (10 March, but the reference to "heat" suggests a date later in the year.

Olga is in a very different position. Hermine and Katti[20] came here yesterday and told me today how well Olga lives, how elegant and splendid her life is, she keeps talking about her rich *uncles*, that's you of course, and a rich *stock market guy* and two others who also *add* their bit.

They say her dresses are all from you in Paris. If I were you, I wouldn't help a woman who has a reputation there for being so vulgar. The girls say she looks frightfully old and ravaged. Not a word about going to *America*, for Hermine knows someone who is well informed about everything concerning Olga. It's none of my business, but considering her behaviour you won't get any respect in Berlin.

The mother of the girls[21] is very sick and yesterday they thought she would die. The news *upset* me although she never did me a good deed, but I feel only good will towards her and do what I can, who knows how I will fare? Who will take care of me? Not even a dog, that's how alone I am and have not a soul who cares for me. During the night I can't sleep for all these thoughts and often cry for hours. The child will be born yellow-faced, they say when one is troubled a great deal during pregnancy, the child will often turn out skinny and very ugly.

How are you, dear Alfred? How long will you stay in Paris and where are you going? I have to stay here in this heat and wish I had relief.

With many kisses,

Your Sofie

SH 8 (8)

Wednesday 8 July [1891]

My dear Alfred,

I won't wait for a reply to my last letter and write these few lines to you because any minute now it will start,[22] and who knows when I'll recover

---

20　For Hermine see Letter SH 6. Katti (Käthe) was another of Sofie's stepsisters. She followed Sofie's example and became the mistress of a wealthy man, the wine merchant Ignaz Weikl (see Sjöman, *Mitt hjärtebarn*, p. 88).

21　I.e., Julie Stern, Sofie's stepmother.

22　I.e., labour pains; Sofie's daughter Margarethe was born on 14 July 1891. She became a language teacher and was working as a volunteer nurse during WWI when she met the physician Dr Ferdinand Böhm. They married in June 1917. In January 1918 Margarethe gave birth to a daughter, Olga. Her husband died of a brain tumour in October 1918.

from it, I am totally run down and that makes me even more anxious, for one needs strength for it, and I've lost all the strength I had. Will I ever recover my looks, the child at any rate will leave much to be desired, for I had troubles and sorrows all through. Perhaps I've caused them myself, because your little toad had bad conscience and that's more terrible than anything else. Ah, if I could be with you, dear Alfred, I would certainly fare better, you can be rather severe, but in many things you are gentle and good, all the others are, to be honest, nasty and selfish as well.

It's only now that I've come to know people and will have a hard time to forget the lesson, perhaps it's better so, or I would have stayed a fool all my life.

Yesterday my stepmother was with me. She looked terrible after her illness,[23] and I felt sorry for her although she was very bad to me and I gave her my last cash so that she can go to the countryside for three weeks, ah, what misery there is in the world! Her children should go to work and sweeten the short time she still has left. How lonely and bitter it is to be without a mother. One feels it only at *certain moments*.

I am so sad because I don't have anyone to be with me, perhaps Bertha[24] will deign to come, but she won't do it from the heart. And now, dear little boy, how are you, are you in good health, and how is your appearance? Dr Barber says he would have visited you but you were unfortunately not in Paris, I knew immediately that you are not fond of him, he is skilled, but not simpatico, frankly speaking.

Write soon, dear Alfred, and greetings from the heart,

Your Sofie

SH 9 (12)

16 Jan 1892

My dear Alfred,

I haven't heard anything from you in a long time and am very worried therefore, because I myself am ill and have no peace, thinking constantly of you and would like to know how you are and what you are doing.

---

23 See SH 7.
24 Sofie's sister. See Letter 147.

You write so rarely and when I do hear from you it is only reproaches and insults that I do not deserve because I have never lived a more retired and economical life, but I would like to pay my former debts which are still in the name of Nobel, because it makes such a bad impression if I don't pay, I promise you never to get in debt again, but Dr Barber thinks one should pay off the debt as soon as possible, or there will be great problems, I beg you many times, dear Alfred, be kind enough to help me, I am in a terrible position. Nor do I have money to pay living costs, and today I had to pawn my last brooch, I have never been in such a bad position as now, I am desperate, I have never been so miserable even at home, and my poor child, what fate is in store for her?

Greetings and heartfelt kisses,

Your Sofie

SH 10 (13)

21 Jan 1892

My dear Alfred,

Many thanks for your dear letter containing 2,000 florins, and for the telegram, you can't imagine my joy on receiving the telegram, dear Alfred. I have so many problems, spend whole nights musing about them, reproach myself and therefore fell ill.

You are the best and most noble man on earth. Your benevolence towards all people is admirable, dear Alfred, and God will reward your benevolence.

I immediately went to Dr Barber and asked him to give you an exact account of my debts, the bills are in his hands. I borrowed the money in your name. The bills aren't in your name. I pawned the jewellery for 1,000 florins, I had to do it because I often didn't have a penny at home. My illness and the care I needed after the birth was very expensive, and every time you sent me money I paid off my debt, so that in a few days I was again without any money to live on, so it went on, and in the end I had to pawn or borrow something, I never knew such trouble and sorrows, it was my own fault because I was stupid and now I have to suffer for it and do penance. Ah, my poor child, who is so beautiful and charming, deserves the greatest pity. I won't talk about myself, I've made my bed and must lie on it.

The letter I received today wasn't very flattering and upset me greatly, day after day I must hear such reproaches.

How are you, dear Alfred, how is your health? Many heartfelt kisses and greetings and again a thousand thanks.

SH 11 (14)

Sunday, 21 Feb 1892

Dear Alfred,

I just received your telegram and am hurrying to send you all the original bills and addresses, I didn't want to do it, because I was offended by the tone in which you wrote to me and about me to *strangers,* so that I said to Barber: "Come what will, even if my life depends on it, I won't take a penny from you and rather go begging."

If you could see me, how unhappy and miserable I am, you wouldn't have had the heart to say such severe and hard words to me. You have a right to be angry, but I am no criminal, and it is the first and last time that I get into debt. I swear it by my child – the dearest thing I have in this world.

You have such a compassionate heart, I never ever expected such expressions from you, such abusive and offensive language, you must have been terribly angry or you would not have written in this form about me to strangers.

It is true that you met me in Baden, but for Heaven's sake, you yourself say that it is no shame to work. Why are you putting me down before all people? And all that because there are bills here in your name. My name is Hess, and everyone knows me by that name, there can be no talk about scandal, once I have the money to pay up, I will never take another penny. I've learned my lesson, I bitterly *repent* it and must *do penance* for it.

Dr Barber will arrange everything and as far as the bill of exchange is concerned, I would not have dared to issue it in your name,[25] that man is in any case a close acquaintance, a business associate of my father's, not a stranger, in fact even related. I needed the money urgently at that time and did not want to write and beg you for it. His address is Obere

---

25 But see her sister's letter of 20 February 1892, p. 283, which alludes to such an action.

Donaustrasse 35, Mr Trensch in Vienna. You can write to him yourself and ask whether I speak the truth about the 2,500 florins. The physicians are very concerned about my health, all those offensive words have affected me so much that I have anemia of the brain and my limbs tremble.

You are driving me to despair, dear Alfred, you force me to commit suicide and then what will happen with my poor child, who has such a *sad future* ahead of her. I tell you again, dear Alfred, you must pardon me and forget all my stupid, frivolous escapades, I was foolish or ignorant but certainly not evil, I don't have any evil in me. You can accuse me of everything else, but not of greed and selfishness, as one usually finds in women.

I send you these lines, begging you not to be so hard on me, I am used to your gentle manner and so this is quite bitter for me, to hear your nasty words.

In former years you were so nice and loving to me, but recently I get to see you so rarely, and when I hear of you, I get nothing but reproaches and sorrow to which I should get used if only you would be with me more often. How are you, dear Alfred, have you recovered your health in Italy?

Here in Vienna, the whole world enjoys itself, only I am wretched and feel lost. If I could only go for a short time to Merano or Bolzano, I am so ill and need care and haven't got a soul here who would show me understanding and affection, I never thought my life could be so miserable. You were so good, my dear little boy, and you wrote that I should give you an account of my debts, which follows, don't take a fright and mainly don't be angry, it will never happen again, I have the greatest problems and sorrows from it. Well, then, my dear Alfred, I will say adieu and beg you to send me at least an advance so I can pay off something.

Many fond kisses and greetings,

Your Sofie

SH 12 (10)

Wednesday evening, [March] 1892

My dear Alfred,

Best, heartfelt thanks for your little letter with the enclosure of 1,000 florins. You are so good and have a heart such as hardly anyone else in the world! How often I remember how well you treated me and your fond benevolence and noble heart, in short, everything I had with and from you. For a year now all that is over, and my only joy in life now is my child, whom I adore and could devour with love. You write nothing about getting together, my dear Alfred. Instead you say I should look for someone else, but I beg you for a meeting, dear little boy, it would make me so happy to see you, you wouldn't recognize me, that's how much I've changed in everything, I am content and simple in every way. Not a hint of pretension, and if I could be with you, my good mood and everything else would return. There are a number of things you want to discuss with me, and that's better done in person, I think, than through strangers all the time – don't you agree with me?

I would have a great deal more to say to you, to ask your advice, which I can't do at such a distance, I could pack up quickly and come to you, even to the end of the world! Of course I would worry about little Gretl, who is so cute and stretches out her hands to me and says "Mama," she is clever, she doesn't have that from me, only the vivaciousness and the perpetual *laughter and friendliness*, she looks like a one-year-old, so big, but her face is very small and delicate and she has wonderfully beautiful eyes, blue, and a charming little mouth with six teeth, that's rare in a child of eight months. Her nurse is good looking and very young.[26] Would you like to see the little one, you would certainly like her. She is my only possession, if I had you too at this moment, dear Alfred, I could embrace the whole world in my mind! But you disagree and are still angry, but I promised you never to go into debt again, and I stand by it, you may believe me, even if you don't *come from Africa*,[27] I live so modestly now and could not make ends meet only because I always had to pay interest and make payments on my debt.

Again, a thousand thanks, and now I beg you from my heart, let me come to you, you won't repent it.

Dear Alfred, I don't even have a brooch to pin on, I would like at least one pawn ticket to redeem it.[28] Please be nice and send it to me.

---

26 An odd remark; is Sofie pandering to Nobel? Sjöman speculates about the erotic appeal to Nobel of a father-daughter relationship (*Mitt hjärtebarn*, pp. 39–41) and quotes an unpublished letter of October 1875 from Nobel to Liedbeck in which he expresses his desire for a "little nursemaid" he had seen in Vienna in crude and explicit terms (p. 21).
27 Presumably meaning "a stranger who will believe anything."
28 See SH 9, in which she laments that she had to pawn her last brooch.

Apropos, just this minute a telegram arrived from you telling me that you will pay Trensch,[29] it's taking a great weight off my chest, the man was unwilling to wait any longer and had already decided to seize my possessions.

A thousand thanks for everything, dear Alfred.

Many fond kisses and greetings,

Your Sofie

SH 13 (15)

Baden, 14 July 1892

My dear Alfred,

Your last letter made me very sad because this is very unpleasant for you. You are so good and benevolent that you don't deserve such a thing. Anyone who is used to your gentle treatment feels so unhappy without you, there can be no talk of [another] love ever. I am very run down, as people tell me, and so nervous that I have to go immediately (that is, on Saturday or Sunday) to Franzensbad. It doesn't bode well for my condition. My child is to be pitied, today she is exactly one year old, and I am as ill as I was then, if not worse. Perhaps the baths and the water will do me good. I beg you, dear Alfred, write soon to me there and perhaps you could make a short trip there. I would be so happy, for my life is awful.

Many kisses and greetings,

Your Sofie

SH 14 (16)

9 Dec 1892

My dear Alfred,

I've received your letter from San Remo with the enclosure of 2,000 francs, and first of all I want to give you a thousand thanks for your

---

29 For Trensch see SH 11. Nobel sent money to pay Trensch (see Letter 200 of 16 March 1892).

benevolence and your dear words give me even more joy, I give you an extra thanks for them because they are so rare, and since I am no longer pampered, those little delicacies taste all the better. I often say to myself in my loneliness: How stupid and simple-minded and unreasonable I was! But it's useless to repent. I woke up to reason too late. I am unhappy and my life is joyless, God only knows where I would be without my child! Believe me, dear Alfred, a little creature like that keeps my mind together (even if it is a small mind) and keeps me from bad thoughts, I should have had such a little piggy a long time ago, then I would have company now, unfortunately the little thing came like lightning into my life, in one way she brought me joy, in another misfortune.

I am too attached to the child, but even so I pamper and care for Bella, I feel sorry for the poor dog, she feels left out, Greterl pats her all day long, but Bella growls and goes away, she likes only her master and me. Now my dear little boy, tell me how you are and how you spend your time in beautiful San Remo, is your laboratory not finished yet? Where do you live, I have no address for you and therefore send my letters to Paris. Will you stay long in San Remo? I wish you a pleasant stay and nice, warm weather.

Here the weather is bad and cold, wind and snow, ah, if I could go south and cure my catarrh, I won't get over my cough easily, it is already chronic and all the inhalations don't help. After Franzensbad, I lost much weight, it's the fault of the baths or the sad times, of which I have many, and what affects me most is that we meet so rarely, when will you come to Vienna, dear Alfred? As for my sisters, I hear only from Malie,[30] but very rarely from America.[31] The children, I hear, all earn money, that's lucky for Bertha, they are supposed to be nice children.

And now, my dear Alfred, I bid you adieu and hope to see you soon in Vienna. Many fond greetings and kisses,

Your Sofie

---

30 Sofie's sister Amalie (Malie) was married to Albert Brunner, director of the imperial zinc works in Celje. Letters from Brunner to Nobel show that they were on cordial terms and that he supported the education of their daughter.

31 I.e., from her sister Bertha Goldmann, who had emigrated to America. See Letters 147 and 209.

SH 15 (17)

Vienna, 21 Dec [1892]

My dear Alfred,

Many thanks for your dear and heartfelt lines and the enclosure of the cheques for 1,000 florins.

I was very distressed hearing that you are not well, as I suspected. When I saw you last you looked youthful and no one would have thought you older than forty-five.

I think that's the consequence of your benevolence, dear Alfred, for no *other* man on earth is as good as you. I wish you all the best and good luck and most pleasant holidays. I would have liked to come to you so that you could see my little girl – she is charming. You would love her prattling, and she laughs all the time, has a sweet attractive mouth, little hands and feet, not like the photograph. The photographer said she was such a vivacious, clever little thing, can you believe it, when he put the black cloth over his head, she yelled all the time "play hide and seek" and ran to him without being embarrassed, she is so cute, more than I can tell you. I don't know from whom she has inherited her intelligence. You, dear Alfred, are the only thing missing to make me happy, and I have lost you through my own fault, I was stupid and must pay for it now. The miserable house *and the third-floor* apartment penalize me as well.

I had to call the doctor today because I suffer from congestions [*sic*], vertigo, and anemia of the brain. He thinks I need fresh air, need to be outdoors as much as possible, but how can one stand being outdoors in this windy weather, it's a terrible climate. How long are you going to stay in San Remo, until New Year's, I suppose. I had a handsome simple purse for calling cards made for you, carry it with you and think sometimes of

Your Sofie. Many fond kisses.

SH 16 (9)

26 Oct [1893]

My dear Alfred,

I thank you only today for your letter with the enclosure of 2,000 florins for the simple reason that I still don't have an apartment and did not even have time to write. I am completely beside myself, so nervous and sick that I can't stand it much longer, I am an unhappy creature without counsel or help from anyone!

You are so good to me, dear little boy, and yet I feel strongly that you take no interest in my fate, of course that's my fault, why was I so foolish and longed so much for Vienna, now I am in a mess and don't know how to get out of it.

I looked for an apartment in Pressburg but found nothing, many people said "later maybe," so I went back to Vienna and am looking now for two months in vain, the apartments are so expensive and all need fixing up, for which I have no money.

You can't expect anything under 2,000 florins and all are on the third floor, too high up for me because I have a *hernia* and can't climb that much, what am I supposed to do with my furniture, everything I saw at that price consists of four rooms, kitchen, and anteroom, and I have seven rooms now, the worst thing are the cockroaches, they are already looking forward to me, they inhabit every vacant apartment.

I'll get very little key money for my place,[32] practically nothing, and today is already the 26th, and I sit here without a roof over my head, a misfortune for myself and the child, I've never been so badly off.

Wherever I move, I would need some six months to arrange things, I can't just move on the spur of the moment, maybe it's best for me to go to [Buda]pest, in every city one has to look around and give careful thought to the location and check out the rest.

In case I find a small apartment, I'll go to a hotel for a few days, because apartments nowadays are in such filthy condition that you can't move in, those people are pigs, let me tell you, dear Alfred, terrible people, I quite believe that Mr Flaum, who took over my apartment, commented on its clean and good condition. Unfortunately I'll end up in a pigsty and have to relinquish my beautiful, elegant apartment to strangers.

I beg you, dear Alfred, write to me and send me the rent money, which has to be paid immediately. *Don't be angry.*

Many kisses and greetings,

Your Sofie

---

32 *Ablösegeld,* paid to renters who voluntarily gave up their lease, usually when this was to the advantage of the owner (i.e. when he could raise the rent).

SH 17 (1)

Tuesday 14 Nov 189[3][33]

My dear Alfred,

Thank you very much for your letter and the enclosure of a cheque for 1,000 florins. Unfortunately it arrived too late. I could find no apartment, they were all too expensive, at the same time I had difficulties because I am not married, these asses here are so prejudiced, and the Jewish landlords are the worst scoundrels, I am completely depressed and in despair because staying in a hotel[34] with a small child during the winter in Vienna means terrible cold and wretched food. I am tired of life because I have no luck, dear Alfred, since I am no longer with you and live a poor life, lonely, in contact with no one and feeling so depressed that I have often thought it would be best to commit suicide.

What am I to do? I can't get an apartment for less than 2,000 florins, but the hotel is even more expensive and absolutely miserable, should I perhaps move to Merano for the winter? It's cheaper there and the climate is nice, if it is alright with you I would go there immediately, write at once, dear Alfred.

You have no idea how a woman suffers in Austria if she is not married. Kapy[35] is willing to marry me, and I am willing in turn because of the child who is so nice and intelligent, you would marvel at her charming prattle, Olga can tell you about how sweet she is. The child would need a name, but how am I to go about it, dear Alfred? Do I have your permission? You are everything in this world to me and therefore I beg you to give me your opinion, how you would like to arrange everything, I am in a desperate [36] situation without advisor. How are you, dear Alfred, how long will you stay in Paris?

So long, fond kisses,

Your loving Sofie

---

33 In ÖI-5 this letter is dated "1890," but the day given (Tuesday, 14 November) suggests the year 1893, as does the context, which resembles that of the preceding letter, SH 16.
34 Presumably the Hotel Tegelhof, which Sofie's father mentions in HH 15 and 16 (undated, in Arkiv ÖI-5).
35 This is the first mention by name of Captain Nikolaus Kapy von Kapivar, the father of Sofie's child, whom she married in September 1895. See Appendix, p. 281.
36 *Desperaten* is misspelled "dasberarten."

SH 18 (19)

5 Jan 1894

My dear Alfred,

A thousand thanks for your benevolence and attention to my little duckie, the doll is lovely and charming, so much so that I'd like to play with her myself, once again, many thanks!

Greterl looked at the doll and got very excited. She calls her the "French girl," you would enjoy hearing her prattling, dear Alfred, I don't want to brag, but she is a whiz kid, you would be surprised how well she draws and declaims. Olga, too, was charmed, in my great sadness and sorrow the child is a beam of light in my life, ah, if I only knew that the poor child will be taken care of and had a name, in my present situation the future is bitter for me and for my child. I must tell you, dear Alfred, that I am very anxious and often think with my head aching, there are moments when I can't keep my thoughts together, and fear I am going mad, I don't know what to do, it must be the consequence of the difficult delivery, don't you think?

With whom can I talk about such things? Only with you, dear Alfred, you are so good to me, you have a heart and a sensitivity that is rarely found in a person, today I spoke with a gentleman who knows you from Carlsbad, he was quite enchanted with you, and there are numerous men like him who admire you, but there are few people who knew you as well as I, and I was so foolish, so stupid and mindless, and a big ass. I acquired the little reason I have too late. How are you, dear little boy, are you well? You must suffer in this great cold, because the Italian stoves are devilish devices, here we have every day minus 14 degrees, you don't see anyone in the street, everyone is swaddled in clothes.

Write soon, and fond greetings,

Your Sofie

SH 19 (18)

Saturday night, 1894

This morning I received your dear letter that contained the money for the annuity,[37] for which I thank you very much and kiss you in spirit for your benevolence. Ah, there is no other man like you, and as long as the world lasts, there will be no other like you. But apart from your noble mind and good will, one doesn't find a man who is as refined as you and as sensitive in his behaviour to a woman, and therefore I am even unhappier than anyone else in my position.

Yes, dear Alfred, today, after a long time I'm at ease again, for I am without worries, yesterday I didn't have a penny in the house, not even anything to pawn. Can you imagine my joy? Once again I thank you many times. I see from your letter that you think the total I gave you is fraudulent, well, my dear Alfred, I can assure you that I got the money from him *without interest*, believe me I don't lie, and I ask you to address yourself to him directly if you don't believe me. Dr Barber is tired of me, I put him off to next week, begged him not to bring shame on me, I would make sure he'd get his money. I ask you, look at his letters and bills, it must be correct because Dr Barber doesn't cheat, he knows that you know what's what, dear Alfred, I couldn't remember the total he paid, I heard of it when he calculated the sum and presented it to you, I beg you check it out. I don't benefit from this, dear Alfred, I only want to live without shame and not stand here, like a beggar woman, without furniture, and therefore I beg you a thousand times, I spend whole nights dreaming of the executer coming and once even that they took away my child, such folly.

Ah, such worries are terrible, I never thought I'd experience this, I went from heaven to hell, and because of pure stupidity. Well, that's how everything comes home to me.

Many greetings and kisses,

Your Sofie

---

37 This is the first mention of a payment being made to her out of an annuity Nobel set up on her behalf, but see SH 22 in which she asks him to raise the amount.

SH 20 (22)

10 July [1894]

Dear Alfred,

Although I am obliged to hear your news from a third party, I cannot keep from telling you that I have converted to your religion and am Protestant now – the child, too, of course – so that we are now closer to each other than ever, and I hope that you will not turn your good noble heart from me. Dear Alfred, you have always been the ideal of my youth, my protector, my friend, and even if love wanes after some time, our heartfelt friendship must endure to the grave. I beg you, remain a good friend also to my child, who is so nice and dear, because I have resigned all interest in life. It is very bitter to have to talk of money, because I am being treated like a common whore, especially by this *vulgar common Jew*[38] who is known in all of Vienna and whom you would find arrogant and a mean dog if I judge you correctly, so I have to tell you that I am quite determined to marry and if you want to pension me off, dear Alfred, I beg you to give me 200,000 florins, then you have the assurance that I and my child are looked after and can live decently, you can invest the money with a bank, so that I get only the interest, it would in any case only be 8,000 florins, how one would have to make economies, considering how I have lived so far, it would mean setting all luxury aside. But I will gladly do it because it is better to be dead than lead a life like this, and I will show up that vulgar Jew. I curse him and his children right to the grave. Today I heard that you don't agree with my travelling to [Buda]pest,[39] I have to go there, I can't marry here, and I can't remain unmarried and live alone with my child, I don't want that, I want to be a decent woman in future and not be exposed to such hangmen and suffer indignities, I am going to show my teeth to that miserable fellow, I suppose you hired him to pay the woman who lived with you for fourteen years, not to insult her, I find that quite ignoble and everyone I tell my story will think the same.

Dear Alfred, you are in a pretty situation, putting yourself into the hands of those men, that can't bring anyone luck to treat a helpless

---

38 Presumably a reference to Julius Heidner. See SH 21.
39 Her future husband's place of residence.

woman like that. Nor do I think it was your intention to have me treated *like a whore*. I cried so much, and God *will avenge me, believe you me!*

What I have suffered over the last three weeks is indescribable! Yes, I have done wrong, but even so I'm not the worst woman or mean, no one can say that who knows me, but to surrender me to that vulgar Jew *that's a sin*, I didn't deserve that, I have wasted my whole youth and must be glad now that the Captain will marry me, although I would deserve a more loving husband. No one wants a wife who has been the mistress of another and lived with him for so long, believe you me. That is why you must be reasonable and not be so hard on me.

I have to stay here in this heatwave and can't get away *because I have no money.*

SH 21 (20)

Vienna, Saturday [summer 1894]

Dear Alfred,

You will have received my two letters from Budapest, I had decided to stay only two days in Vienna and go immediately to Merano to recover my health, but unfortunately I was here only a few days before I once again fell very ill and am laid up now at the hotel in two dark and nasty rooms, where I have neither air nor care.

The result is that Dr Breus, who delivered the child, will have to operate on me.

I have suffered a great deal and look like a corpse, I can't look after myself because I have no money and must beg for every penny.

Professor Breus is very sorry for me and takes a great interest in my fate, because all my suffering is the result of the abuse I suffered at that time from that *dear Jew* Sternlicht, the half-Jew Philip[p], and the lousy mercantile soul, Mr Heidner,[40] who treats me like a whore.

---

40 Leopold Sternlicht, Maximilian Philipp, and Julius Heidner were lawyers and on the board of the Nobel Dynamite Trust Company. Heidner was appointed trustee on 3 July 1894 and supervised the dispensing of the annuity Nobel had granted Sofie. See the official notice placed in the *Amtsblatt zur Wiener Zeitung* of 10 July 1894, which gives as the reason *Verschwendung*, wastefulness or extravagance.

Tell me, dear Alfred, don't you have bad conscience, considering how good you are to all people, I believe you don't know how ill I am or I am convinced you would look into this yourself and treat me differently. My health is completely ruined and I suffer terribly, for three months now I have such bleeding that surgery is necessary and immediately, I am afraid of the anesthetic and keep putting off the surgery, because if anything happens to me, what would the poor child do? What does Heidner know about feelings? I hear he treats his wife like an animal, he is as much suited to be my trustee as I am to be a tightrope walker, he is a heartless fellow, how such a creature can become a director is a puzzle to me, well, society needs people like him, but I won't take such abuse from that mercantile soul, and one day the world will hear of me, I care little about life, I am fed up living a life like a dog, it's not enough that you don't give me enough to survive, I have to beg for the money and am exposed to scandal, for example, the headwaiter at the hotel made a fuss because I didn't immediately pay, so that a gentleman in the hotel had pity on me and couldn't understand why I am in this *miserable position*. You happen to know that gentleman, a Frenchman, since that time I am so ill and nervous, that I jump at every sound. You wouldn't recognize me the way I look now, and because the physicians insist on an operation, I would have to go to a hospital for eight days. That's the position I'm in since I have *no home*, dear Alfred, the whole world scorns me because I am not married and have a child, you know what the Austrians are like, a stupid people. If you don't have a marriage certificate tacked to your back, they believe you are a common whore. The married women sell themselves to the first-comer for a dress or a hat and cheat on their husbands, but a minister or a priest has presided over their wedding. Therefore I am determined *to give a name to my poor child* and to marry her father, I will make that sacrifice and may not be as happy as I deserve, because without wanting to flatter myself, I am a good mother and sacrifice much for my child, *God only knows*. If I wasn't so scrupulous, I would certainly be better off, and I wouldn't have to rely on charity.

You are not fair, dear Alfred, if you thought how I spent my youth without the things other girls enjoy, you would be more feeling and have pity with me.

You have such a good heart as hardly anyone else, therefore, dear Alfred, I beg you to deliver me from this wretched life, I can't stand it any longer. In a few days, when I feel a little stronger, I will go to

Merano to recover my health, because I am totally sick in body and
soul, no one in the world can imagine what I suffer. I beg you arrange
for me to get my money monthly, without depending on *that gentleman.*
I never thought, dear Alfred, that you would expose me to such misery,
Dr Sternlicht has earned 4–5000 florins or more in your employ, I know
it from *a very good source.*

Many greetings,

Your Sofie

SH 22 (21)

[1894/5]

Dear Alfred,

I have been ill now for three weeks with peritonitis. Dr Ellischer is treat-
ing me. My suffering is the result of those bad times and the upset I
went through recently in Vienna. I am so worn out that I can't walk
by myself and of course I can't take care of myself because I have no
money.

I don't know what to do without money or support, I live here in a
cheap hotel, wanted to stay only two days, but suddenly fell ill, the heat
in the room is almost 40 degrees, it is to die.

I wanted to go to the mountains for four weeks or to the sea to recover my health, and will then settle in Carlsbad. The Professor thinks
I am a giant if I can keep on living with such bowels. I don't think you
want me to commit suicide and bring shame on you, you are so good
and noble to everyone, and you treat me of all people in this manner?
I know you don't believe in any God, but there is something to which
you are attached in life, namely the knowledge of having been fair. But
you are not fair to me, on account of that lousy debt, which you could
have paid with *half the money,* and you push me into misery, *cursed
be that rotten Jew,* you will hear from him, I am sure that scoundrel
has packed away a neat little sum, I know him, and Barber's family
too, his father was the worst usurer in Vienna, and yet he is your lawyer, Philipp protects him, because he too is a baptized Jew, and many
people in Vienna know who he was and how he made *his money.* I
could have *made up to* that old Jew a long time ago, and he would have

pimped me too,[41] but a creature like him doesn't suit me, *let him work his daughters to death*, so that Vienna has two miscreants fewer. Those friendships *of the ladies*! If I was alone with you, I would tell you a few stories, *your hair would stand on end, they are all scoundrels like Singer, not a hairbreadth better*.

Both of them, he and Philipp, want to offload their daughters on you and are mouthing off because you have *Miss Frohner*[42] as a lover now, let me just tell you *one word*, and you won't associate with them again, I am convinced, but I will keep silence until the right time comes, God who is *my witness* knows I won't keep silent about what I have suffered in Vienna, which has made me sick.

When I left Vienna, I got 500 florins and had to pay the hotel and the tips for nine months, which amounted to 140 florins, and the travel costs, luggage, and now my illness, the doctors, food, all of that is supposed to come out of that sum?

Heidner begged me to take the money, I didn't want to accept it, I'd rather go begging than take his alms, God will not desert me, because I have always acted scrupulously and well without considering my interests, that's why I am in this poor condition now and you had me during my youth, when I could have met rich men.

Now you discard me because you have others and prefer them, are you not afraid to act in this sinful manner?

I am in such a terrible position that you must expect the worst.

Many greetings,

Your Sofie

---

41 Amalie Brunner, writing to Nobel from Celje (20 February 1892; see p. 283) refers to Barber as Sofie's *Hausfreund*, her "special friend." The word sometimes refers to a lover, but seems to have a more sinister meaning here, since Amalie asks Nobel not to tell Barber anything about Sofie's horrendous debts because this might enable him to blackmail her.

42 Not identified; perhaps a relative of the Johann Frohner, a Viennese millionaire, who ran the Hotel Imperial.

SH 23 (23)

Merano, 22 Jan 1895

Dear Alfred,

I am so miserable and so desperate that I must beg you once more to come to my aid. You are so good and have a kind heart, why are you so hard and severe on me? When I read through the letters you wrote me years ago, tears come to my eyes, and I feel so unhappy and abandoned, I acted without thought and without reason. Ah, you were so loving and noble and good to me, and now I am sick, not a single person cares for me – unmarried with a child and without means or protection. I don't have a penny, have been for days now without cash, I beg you help me, I sent you a telegraph today, I didn't know what to do anymore, because I get money from Heidner only on the 15th of next month, I have to make many payments here and don't have a penny to do it. Send me something, dear Alfred, I urgently need it.

Many greetings,

Your Sofie

SH 24 (39)

Merano, Sunday 10 March [1895]

Dear Alfred,

Don't be angry that I come to you once again with a request, but I have no money, am poor like a dormouse. The amount I will receive on the 15th will be needed for the rent and for food. I don't have enough money even to buy a dress for the spring. I have to economize and dress very simply. People who know me from before can't believe that it is me, and yet it is so, dear Alfred, here I am all alone and without means, you are my only support and I thank you from my whole heart, I alone know how good you are, dear Alfred, and what I have lost through my stupidity. I keep repeating that there is no man in this world who is as good and sensitive as you are, dear Alfred. I think of you so often and very much and have read something very original, which I send you, to make you laugh.

How is your health, dear Alfred, is the winter there as severe as here? How is Olga? I still have no reply from here, where is she now?

I beg you many times, dear Alfred, be kind enough to help me with a bit of money, I need it so urgently for the spring, and where am I to get it?

Don't be angry. Heartfelt greetings,

Your Sofie

SH 25 (24)

Merano 22 March [1895]

Dear Alfred

From my trustee, Mr Heidner, I found out that you have no objection to my marrying Captain Kapy. I take this step for the child's sake, whom I love above all else and to whom I wish to give a name, so that she wouldn't suffer shame later on, and that people wouldn't point a finger at her, the poor child, for being illegitimate. If the mother erred, why should the poor creature suffer for it all her life?

I know and have experienced the suffering, dear Alfred, and still suffer because I am not married, it is terrible, let me tell you! I am often in tears because I have a tender heart and am sensitive and feel keenly how they treat an unmarried woman with a child in Austria.[43] My child is beautiful and delightful, everyone is charmed by her, only you, my dear Alfred, pronounce judgment that you cannot seriously mean. I was quite hurt and sick when I heard from Heidner what you wrote about my beautiful child, although you promised me in the Ebendorferstrasse[44] to take care of her, if the baby is a girl, and not to be angry. You wrote the illegitimate child is none of your business and to make out the insurance such that she would get nothing after my death.

No, you can't be so cruel, dear Alfred, I consider you too noble and good. Really, you will fix the income I receive for life in the way

43 Yet Austria had one of the highest rates of illegitimate children – 27.8 percent in 1870 according to the *Routledge History of Women*, p. 24.
44 Sofie's address when she lived in "palatial quarters"? See Letters 182 and 207.

I suggested to Heidner, so that my child will get the money after I die, won't you? I am in any case in such poor health and constantly ill, how long will I live – and my child is so young – your heirs surely won't need the money that would revert to them if it is an income for life. Heidner wrote it will all go back to your heirs after my death, and my poor child would have a miserable future. As far as the insurance for my child is concerned, I can't take anything out of the 500 florins,[45] I can hardly make ends meet as it is, I can't even have a dress made for myself, I am wearing a winter dress, my dear Alfred, have no money even to have the simplest dress made, and the shoes from Nevaux are stuck for more than two months now in customs and I can't redeem them, sometimes I don't have even a gulden at home, I have never been in such a miserable position. In addition, I am sick and so nervous that I look ill. How happy was my life years ago, and now?

After living with you, dear Alfred, it is difficult to find happiness with someone else, that is the greatest punishment for me, believe you me, dear Alfred.

I beg you be kind enough to send me something. I am totally without money and need to buy so many things, because it is already quite warm here.

With heartfelt greetings,

Your Sofie

SH 26 (40)

Merano, 4 April [1895]

Dear Alfred,

I would not have sent you a telegram if I weren't so desperate. Consider that I am stuck without a penny here in Merano, where everyone knows me and many people believe that I am still doing as well as when I was with you.

I don't know what to do, dear Alfred, and I beg you again to send me some money, you can well imagine my situation, to be stuck here without money, to have no one to help me, such a life is bitter and very

---

45 The monthly payment she received from her annuity.

miserable for a person who has seen such good times with you and because of you.

I am disconsolate and can't understand why you are leaving me in such a nasty position, you have such a good heart and are so loving and gentle to your friends, and so hard on me.

I certainly wouldn't approach you if I had any means, and consider, dear Alfred, that I can't make ends meet even with all my economizing, everything here is expensive and especially now when there are so many foreign visitors, often I don't even have anything to eat, I am sure you would help me, if only I could convince you of how I economize by comparison with before, but I have to eat, for the rest I go without – no dresses, no amusements of any sort. Please, dear Alfred, be kind enough to help me with some money. I only get my annuity payment on the 15th. Don't be angry. Heartfelt greetings,

Your Sofie

SH 27 (25)

Merano, 6 April [1895]

Dear Alfred,

Many thanks for your kindness, you are so good, I have no more words to thank you.

The 800 marks freed me from the greatest worries, by God, I didn't know how to go on, you are and will always be the most benevolent man on earth, one things hurts me terribly, that you sent the money incognito, a man should be more conciliatory, but you, dear Alfred, seem to forget that we all have faults, I have some perhaps, but at the bottom of my heart I am good and grateful, dear Alfred!

What I needed most urgently were the shoes from Neveux-Senne, they were waiting here for an eternity and now he had them returned to Paris. By God, I don't know why Olga didn't write to me for such a long time, but all women are the same.

I must thank you once more, dear Alfred, because I was in a terrible position and totally downcast, you are my guardian angel in my deepest desperation, dear Alfred, as I wrote to you, I have a bronchial catarrh now, and indeed a very serious one, can you believe I didn't realize it, I had a terrible cough, already starting in November, and only now a doctor saw me and could hardly believe that I did nothing for so long. Now I have to take a treatment that consists of cross bandages and inhalations before I go to bed. The bandages would be very good for you, dear Alfred, if you still suffer from bronchitis, I believe the warm climate will cure you. How is your health?

I received a letter from my poor sister Bertha, she is doing very badly in America.[46] Imagine, she wrote to my father that they live on tea for days and have no penny at home, and poor Bertha is always sick, she can't stand the climate. What misery, unfortunately I can't help them, if I had saved my money, it would be different. Now I understand that, but unfortunately too late.

I sent poor Bertha a few little things, but it is so little, I beg you, since you have such a kind heart and have done so much *for the poor* woman, remember her, it would be a good deed, you can't do anything better, dear Alfred, and God will pay you back twice. Here is her address, I wept a great deal when I received her letter from my father, she did not want to write about this to me, since I no longer have the means I once had. Poor Bertha, I am told, has aged so much from being constantly sick.

Again, many thanks, and heartfelt greetings,

Your Sofie

SH 28 (26)

Merano, 2 May [1895]

Much beloved Alfred!

Don't be angry that I once again ask for your help, you are the only one I can turn to because I have no other support and you, dear Alfred, are the only one who can advise and help me!

I have to move out on the 15th of May and don't know what to do next, I have no means, in short no assistance.

The journey back to Vienna costs a great deal of money, the luggage, the rent on 15 May, in short I don't know what to do and beg you to help me.

---

46 For Bertha see SH 8.

I have such a bad bronchial catarrh, a terrible cough, I should go to Reichenhall[47] to be treated there, here the air is too dry and my cough gets worse, I am completely run down, and I suffer from night sweat, just like you, a disgusting illness, I am practically a cripple, I miss the good life and the kind way you treated me, my dear Alfred, and even if I had a great deal of money no one could replace you, there is only one man for me in this world, and that is you. I have to tell you that, although you treat me very badly now and if you weren't so angry concerning my debts, you would not act like this, that isn't your habit, to drive a woman into misery with whom you have been together for so many years, so be gentle and good to me, I am left here abandoned and without means. What shall I do next?

I am no longer young, and I stupidly wasted the fair time of my life and regret it, my dear Alfred, what is left now but to burden you since I have not saved up even a penny, and I was never selfish and can never be.

What would the position of a French woman be? Very different from mine, a woman with a child, you must consider as well, how nicely I looked after you and without any self-interest, Alfred, therefore don't be hard on me and don't let me suffer so. What hurts me most is your *unwillingness to write*!

Nor do I know what your plans are concerning my marriage to Kapy, I suffer a great deal on that account, people in Austria are such asses and believe that an unmarried woman with a child is a criminal.

That is why I want that torment to stop, they point fingers at me and I am insulted by the domestics, I tell you it is terrible, therefore I wish you'd tell me your opinion at last, but the divorce from his[48] wife costs 500 florins, otherwise he can't marry me.

I beg you, dear Alfred, write to Heidner and give him instructions and for myself I beg you send me something so that I can depart from here, or I'll have the greatest difficulties. I am stuck without means, help me, dear Alfred, I thank you in advance.

Many greetings,

Your Sofie

---

47 Near Salzburg. Letters 30 (28), 31 (29), and 32 (30) are written from Reichenhall.
48 I.e., Nikolaus Kapy's.

SH 29 (27)

Merano, 8 May [1895]

Dear Alfred,

I don't know how to help myself and beg you once again to help me out of this terrible situation, don't be angry, but I am desperate, I am stuck here helpless and abandoned, I must move out of here on the 15th and have not a penny to manage it. I don't have the money to travel, no dress, no shoes, in short I am as poor as a dormouse, imagine, the shoe-maker[49] wants me to pay up front before he sends me another pair because they were returned to him twice already and he had to pay for it.

You can't imagine how it is, dear Alfred, in spite of making econo-mies I can't make ends meet, now suddenly I have to live like a poor woman, from 40,000 I am down to nothing. You were always so good to me and the way you treat me now makes me completely ill, you have pampered me so that I can hardly believe there are other types of men. Yes, unfortunately there are men who are *very different* from you, and don't ask in what way!

I suffer a great deal, dear Alfred, I have a terrible bronchial catarrh and an awful cough, what can I do to get rid of it? The advice is to go to Reichenhall, but how do I get there and will I live there with troubles and worries, as I live here?

Ah, it is an awful life, to be always and constantly in such a position, you have no idea what it's like.

I can't stand this life much longer, my nerves are too weak.

I beg you, dear Alfred, deliver me from the torments here, don't be so angry.

Heartfelt greetings,

Your Sofie

---

49 See SH 27.

SH 30 (28)

Reichenhall, 1895

Dear Alfred,

Once again I come to you, begging you to help me out of this terrible situation. I can't stay here because I am without means, a stranger in this city where everything is so expensive, consider what you are about, leaving me in such a miserable situation. You are good and help everyone, and maltreat only me. I don't know what to do next. Heidner has no idea of the cost of treatment and is too heartless to understand that I must live with my child.

There are days when I don't even have the money to buy us dinner, you can't be so hard, I beg you send me at least something so that I am not stuck so completely without means. Where am I supposed to get any money, dear Alfred?

I am going from one illness to the next, for the past four days I have periostitis, misfortune pursues me in a terrible fashion, ever since I lost you.

You yourself know how awful it is to be sick, and then without means.

Many greetings from your unhappy Sofie

SH 31 (29)

Reichenhall, 15 July 1895

I have heard from Heidner that you have been kind enough to let me have 500 florins,[50] for which I give you many thanks, but I have seen very little of the money that I need so urgently because Heidner sent only 200 florins and the rest, as he writes, will be used for other purposes. You know, dear Alfred, when one undergoes a treatment, one needs more money, and especially here in Reichenhall where everything is so expensive and bad. I am staying here in a miserable room, am very sick, dear Alfred, and can consider myself lucky to be alive, I am not very keen on life, but what would my poor child do?

---

50 Nevertheless Nikolaus Kapy, Sofie's future husband, wrote another begging letter to Nobel on 20 July 1895 (see Appendix, p. 281).

I labour under a bronchial catarrh, was in bed for 2 weeks and have such a cough as terrible as I had some years ago, I constantly have to do inhalations and drink whey, it is an awful sickness, let me tell you, dear Alfred, and how it gets me down, I am practically a skeleton. You are lucky you got rid of this bothersome illness, the worst about it is the *perspiration*.

I *beg you be kind enough to write to Heidner* and tell him to send me the 300 florins, which would rescue me from the worst embarrassment. You have no idea how bitter it is to be sick and without means as well, you know, dear Alfred, what everything costs and if one has to save money to that extent, one can't undergo treatments, and is just anxious. Be kind enough to help me, Heidner is heartless, but not to himself, that's a different matter, there are few people on earth like you, but I don't know why you are so hard towards me. Take pity on me, I have enough to suffer. How is your health, dear Alfred? How is Olga?[51]

Many greetings from your Sofie.
Hotel Deutscher Kaiser[52]

SH 32 (30)

Reichenhall, 23 August 1895

Dear Alfred,

I never thought you would be so heartless, to leave a woman with whom you have lived for so many years in such misery and sorrow, it is incredible. You are unjust and one day you will regret it, believe you me, dear Alfred. I am in a position too awful to describe and I am constantly sick, now I have kidney troubles as well, and no money to undergo treatment. Here I am and can do nothing, because Heidner keeps me so short that I can barely manage to get dinner, is that fair? Every month he deducts 100 florins, and I am supposed to live on 400 florins and undergo treatment, when it is so expensive, please take pity on me

---

51 Olga appears to be residing in Paris at this time. Nobel's will, dated 27 November 1895 and leaving her 100,000 francs, gives her address as 10 Rue St. Florentin, Paris.
52 The hotel, a modest inn and brewery, was built in 1834. In 1896 it was taken over by a Bavarian hotelier, who rebuilt and expanded it into a splendid complex.

and don't make me desperate, help me, I write these lines from my bed because I am very sick and completely run down in body and soul.

How are you, dear Alfred?

Why don't you permit Olga *to write*?

Your Sofie

SH 33 (31)

Vienna, 17 Sept 1895

Dear Alfred,

I have been in Vienna for a few days now, staying in the Hotel Continental[53] because it is cheaper there, and I remember with sadness and an aching heart the good times when I spent such wonderful days with you, now everything has come to an end, dear Alfred, I see a future that makes me completely unhappy, but it must be, and so I submit to my fate. You know and have known it all the time: I marry to get off your back, but I must ask you one thing, dear Alfred, give me a small trousseau, that doesn't cost much, you know I have never had much linen, and now hardly anything is left, send it to me as a wedding present that I may get for myself two dresses and linen, how can I live on 400 florins and acquire all that, you have always been good and kind, don't be so hard to me now that everything is coming to an end, don't forget me. I get nothing of my 400 florins from Heidner without begging. The wedding costs a great deal of money, please consider that, and send me something. I *beg you many times*. How are you, dear Alfred, how is your appearance? Don't forget me. My heartfelt greetings,

Your Sofie

---

53 At 7 Praterstrasse, close to Heinrich Hess' place of residence at 50 Praterstrasse. The hotel was established in 1873 and is now the site of the Hotel Sofitel.

SH 34 (32)

Vienna, Wednesday 25 Sept 1895

Dear Alfred,

My wedding will take place on Sunday in [Buda]pest, and I would be happy to see you there, if it is possible for you to come and bring me luck, dear Alfred. I will stay there for two or three days, then return to Vienna to see that my furniture and other things *I have* in Vienna will be transported to Merano.

I asked you to give me 10,000 instead of 6,000 florins, surely you won't refuse me because I have to look after a child as well and don't know how long I will live because I'm always ill and now have kidney troubles and at my age one never knows what can happen. I am no longer young and the poor child is without means, since the Captain has no money nor possessions and very little heart.

Be noble and give something to my poor child, who is so clever and dear and *shouldn't be made* unhappy on account of me, I beg you do something for her, it would be a *good deed*, dear Alfred, if you do it. I also begged you to send me a sum to buy linen and a few pieces of clothing, I have nothing to wear and no means to buy anything. Heidner isn't a man who will show consideration, he has no heart and no feelings, it is bitter to be dependent on a man like him, one has to beg for money.

How miserable and poorly I live now, perhaps the time will come when you will admit your mistake, you have no woman friends left (and that includes Olga) who will look after you and will be good to you the way I was, completely unselfish to the maximum, a woman like that you will never find again, believe you me.

And now adieu, I wish you much luck and beg you not to forget me and the child.

Sofie

SH 35 (33)

Budapest, Thursday [Oct 1895]

Dear Alfred,

For the past ten days I have been very sick and write these lines from my bed. I had an embolism as a result of all the excitement and worries, and the doctors don't know how to help me. I must thank you very much for your telegram and beg you, dear Alfred, to help me with some money because I am lying here in a hotel and have costs and expenses. I am so weak that I can no longer sit up, my only wish is to get to Vienna and have *surgery* immediately, or I can't get better. I beg you once again, dear Alfred, set aside all anger now that I am married and consider that I have become very unhappy and have made great sacrifices for my child.

Many greetings and kisses,

Your Sofie

SH 36 (34)

Vienna, Friday evening, November 1895

Dear Alfred,

I returned from [Buda]pest sick and weak and had to go to bed immediately because I developed a strong bronchial catarrh.

I suffer a great deal on that account, my doctor says it has become chronic, and I will become asthmatic if I do not take care to cure it as soon as possible. The doctor counselled me to go to San Remo or Abazio.

It is a misfortune to be so sick and have no means, I am sure that you, dear Alfred, will give me the means to take a cure, because what would the poor child do without me, her father is not the man to take care of her, I only took that step to give the child a name. Otherwise I would not have done it for all the money in the world, nor will I stay with him because I can't stand such a life and all doctors who treat me advised me to leave him.[54] My illness is the consequence of his treatment, especially after a man like you, it is quite impossible to live together with him. Mr Heidner of course found it quite alright, but why? Because he himself maltreats his wife and was born without feeling or heart, that is how the matter stands, my dear Alfred.

---

54 But it seems that it was Kapy who left his wife. They are last documented together on 4 January 1896 when he addressed a letter to Nobel from the Hotel Continental, confirming that Sofie signed a contract, which set conditions for paying off her debts and returning to her some of her pawned jewellery. SH 39 (tentatively dated February 1896) indicates that the couple separated after that transaction. Kapy died in 1913 at the age of fifty-seven.

Dr Lang at the Hotel Continental, where I am staying now and who treats me and knows your brother Robert, says I must go to a sanatorium and undergo surgery, it was very *important*, and my health depended on it. Mr Heidner wants me to go to a hospital and stay there for two weeks. Dr Lang did not want to accept this proposal and said he could not believe that you agree that I should go there. I know that you wouldn't want me to go there to save money, dear Alfred. Mr Heidner is very moody and very rough, maybe he has commercial training, but he is quite without finesse.

I have to stay here for another eight to ten days, and Heidner does not want to pay for the hotel. Where am I supposed to get the money? From the 400 florins, I can't even afford a dress to wear, everything goes to the physicians and the apothecaries. It is a misfortune to be ailing. How happy I was with you, dear Alfred, I miss that life and the way you treated me, and everyone is astonished to see what I look like now by comparison with then, I won't be able to go on much longer, because I have a weak constitution, which improved only when I was with you, in your hands, when I knew nothing of worries, if I did not have such a dear child I would have taken my life long ago. I beg you, dear Alfred, send me a small sum so that I can finally undergo the surgery and take care of myself, I thank you many times for all your kindness and benevolence. With many greetings,

Your Sofie

SH 37 (36)

Vienna, 6 December 1895

Dear Alfred,

Although the three greatest men of Vienna, *Philip[p]*, *Heidner*, *Sternlicht*[55] – I call them *the oracle of Vienna* – disapprove of my writing to you, I do it nevertheless and thank you first of all that you are so kind to me and have given me an annuity. It is small, and I have to economize to make ends meet. It will only be possible in Merano, and my health will perhaps improve there as well, for as you know I am

55 See SH 21.

continually sick, I write these lines in bed, am again suffering from my old ailment, and will have to undergo surgery in spite of my fears. I beg you to redeem my earrings, the only thing I still have, it isn't much, just 1,000 florins, or else they go to the pawnbroker, and I am wearing false pearls, a shame, Heidner is keeping all my jewellery and will sell it for nothing, that is really terrible, no one can believe it, I am treated in an unbelievably rough and nasty manner by this devil, and how *can it be different* if you yourself write such letters to him, which of course he reads out to me.

That's not nice of you, after we lived together for 18 [*sic*] years, writing such words sullies your own hands. Perhaps you will come to realize one day how unfair you were to me. I am not the worst woman, dear Alfred, and *you are wrong* to think that my successor[56] is better, any nicety, anything good in her, she has learned from me, you must admit as much. Olga would have taught me differently, the way she told me to treat old men. You did wrong to show yourself with her so publicly in Carlsbad, the world was scandalized, not because she is young, no, they thought she was pregnant, that is how fat and old she looks, you can say what you want, dear Alfred, but she doesn't compare with my elegant manners and appearance, several people said so here, who saw you there. Perhaps something will happen to Olga, God does not leave such behaviour *unpunished*, the way she treated her own *mother*, and she hasn't congratulated me or written, does nothing, the punishment from above will come. I could tell you about the telegrams she got here in Vienna from men in Hamburg, Berlin, Lübeck, everywhere she has lovers, and you go and show yourself with such a person?

Philipp knows much more, I will write in more detail if you are interested in what he tells me. But enough of that, I beg you to keep listening to me. A few days ago Heidner and Sternlicht informed me about the contract taking care of me and asked me to sign it.[57] Since this is important business and I don't understand about such matters, I turned to Dr Elias, my former teacher and now a lawyer, who also represented me

---

56 I.e., Olga
57 In a letter of 30 October 1895 to Heidner, Nobel enclosed a cheque for 6,000 florins with instructions to set up an annuity for Sofie, "under the condition that I will hear no more from Mrs K. and will ignore her letters" (unpublished letter cited by Sjöman, *Mitt hjärtebarn*, p. 81). She signed the agreement concerning her annuity, as Kapy confirms in a letter to Nobel (see p. 282).

in Döbling some time ago,[58] and allowed him to read the document at Dr Sternlicht's office. In Dr Elias's opinion, the contract is unfavourable to me and risky, it does not take care of me because it could easily be that someone to whom I owe a sum sues me and I lose my settlement. In Dr Elias's opinion the contract should be phrased without *qualifications*, merely stating that I will be paid from the said bank 6,000 florins as long as I live, I will pay my debts as I wish and it has no bearing on my annuity, I commit not to burden the annuity with debt, that is, can't borrow on it, otherwise the annuity would revert to the owner. The contract must be notarized or it won't be legal.

That is what Dr Elias said, I can't sign a contract in another form, if I want to be sure to be taken care of, your intention, dear Alfred, is no doubt the best and most noble, and you will find it fair when I use the intervention of a lawyer friend, it seemed to me that I am too stupid to deal with this important question concerning my life, and Dr Sternlicht is for certain reasons not very friendly towards me (a question of vanity), the Jews are *very egoistical*.

I beg you to give me your consent so that I can depart for Merano because I am ailing all the time and could die from one day to the next.

Many greetings,

Your Sofie

SH 38 (37)

30 December [18]95

Dear Alfred,

My most heartfelt and fondest wishes for the New Year. At the same time I thank you from my heart for your kindness and benevolence, but I will never forget that you didn't answer my last letter.

Many greetings from your Sofie

---

58 I.e., in the purchase of the villa in Döbling. See Letter 139.

SH 39 (35)

Vienna, Wednesday [February 1896]

Dear Alfred,

Although you don't answer any of my letters,[59] I come begging to you once again, help me to get out of Vienna.

Heidner, who is in charge of my money, has all of my jewellery, the *two sets of earrings, a brooch, two rings, two bracelets*. The other things he sold, but I think he bought them all for his wife, so many things for 500 florins, a sin. I don't even have a brooch, am wearing an imitation for 2 florins and earrings from Paris, also imitation. I gave him a pawn ticket for everything he has, he redeemed it, and now he wants money from me for it, but where am I to get it? And that is a trustee? He allows his servant, an old fellow, to treat me in a rough and nasty manner, I don't understand his actions and will simply hand the matter over to a lawyer, I don't have to take such insults from an employee.

That philistine talks of you in the same vulgar manner, and if I have an opportunity to see you, I will tell you and open your eyes to the real Heidner. But I am in the greatest difficulties and can't get out of Vienna, my decision is to move to Wiesbaden and Heidner doesn't want to let me leave Vienna and doesn't give me any money for the move, in short he doesn't want to do anything and yet he is in charge, I and other people as well can't understand it, tell me what I am supposed to do? Heidner says if you send him some money, I can move, I can't stay any longer in the hotel, the people are worried because Heidner keeps saying things about throwing me out and other such nasty statements. I am fed up with this man and will change my tune, I won't be treated like that.

You obliged me to marry Kapy, he is gone for a whole month now, and takes no thought for the child, a fine gentleman, he wanted 50 florins every month, and I don't have the money, because with those few gulden I can't live in Vienna. So he left and is running around with women, I only took this step in obedience to you, dear Alfred, and now

---

59 On 30 October 1895, Nobel informed his lawyer Julius Heidner: "Now that the [financial] arrangements have been made, it is my wish and desire never again to hear of Mrs Kapy and to ignore her letters" (letter in the possession of Olga Bëhm, cited by Sjöman, *Mitt hjärtebarn*, p. 82).

you must be kind enough to help me, I am stuck here, quite desperate, without any support, alone with my child. My father is useless, and I don't have anyone else in this world, I beg you to help me and please do so before it is too late and I have to use desperate means.

I beg you help me once more, dear Alfred, you got me used to a luxurious life, and now I have ended up in such poverty.

Your Sofie

SH 40 (38)

Vienna, 16 Feb 1896

Hotel Continental[60]
Dear Alfred,

I write to beg you before I depart for Merano to send me a small sum so that I can look after myself. I am ailing all the time and ailing so seriously that I am forced to have the attendance of a physician and that is quite impossible on 500 florins, I can hardly buy the most necessary things. Be kind enough to help me this once with a little money, you see I don't bother you, I'd rather go hungry than address you,[61] you would be surprised how I live and how things have turned out for me, I beg you be kind enough and send me a little bit. I thank you in advance and send you many heartfelt greetings,

Sofie

SH 41 (41)

Tuesday [February1896]

Dear Alfred,

As you will know, I am still at the Hotel Continental, I have no money and they won't let me go unless I pay the small bill of 250 florins. But

---

60 See SH 33.
61 This must mean that she sent this letter through Heidner or another intermediary.

to come to the point: one of the administrators came to me yesterday and asked me to write to you. He wants a patent concerning a machine for cleaning[62] carpets, a patent you took out in Paris and sold. Here in Vienna, the gentleman says, no one has taken out a patent, and so he would like to get it from you if it is inexpensive or if you want to give it to him since it won't be of any use to you. This gentleman has developed the same invention and when he wanted to patent it he found that Mr Nobel had already patented it in Paris.

My poor father could obtain a bit of employment there as a manager, I can't give him anything in any case since I don't have anything to live on myself.

I beg you, reply to this gentleman by the name of Meyer, he is keen to get the patent from you. He said that Portois & Fix,[63] where my furniture is stored, did not want to buy the patent, but this company will.

Greetings,

Your Sofie

The gentleman's name is Meyer, Hotel Continental. He lives in *Rasau*.

---

62 *Tepl[p]ich Blaserei*. Vacuum – or steam cleaning? The list of Nobel's patents contains one for "an apparatus for the production of steam" (see http://www.nobelprize.org/alfred_nobel/biographical/patents.html).

63 An elegant furniture and interior decorating store, established in 1842. From 1892 on the owners also offered storage facilities.

Appendix

The file Nobels Arkiv ÖI-5 in the Swedish National Archive contains, apart from the letters exchanged between Nobel and Sofie Hess, correspondence between Nobel and members of Sofie's family. I include a selection of letters from Heinrich Hess, Sofie's father; Amalie Brunner, her sister; and Nikolaus Kapy, her husband. They provide significant background information and help put the correspondence between Nobel and Sofie Hess in perspective.

A

LETTERS FROM HEINRICH HESS

Vienna, 18 May 1887

Respected Mr Nobel,

I assume you will not be offended if I take the liberty to send these lines to you today. Coincidence is an important factor in life, and I may state today that it was nothing but a happy coincidence, respected Mr Nobel, that my daughter Sofie made your acquaintance ten years ago. She was inexperienced and thoughtless – another man might easily have deceived her and introduced her to bad company. It was a most happy coincidence, indeed God's will, that my daughter Sofie found such a kind, principled, and well-respected man. The proof is in the fact that even now you still protect and care for my daughter, respected Mr Nobel.

I want to keep my letter short and offer an explanation on behalf of my dear Sofie. I lost some money in the Steiermark[1] and had a large family when Sofie left home. She was forced by circumstances, and also by the fact that her stepmother[2] was not as gentle and kind as she should have been.

Sofie was supposed to make her own way. She was a good and kind daughter to me, and still is, as is shown by the fact that throughout those ten years she always stayed in contact with us. Sofie supported me throughout that time. Since my income is very small – I am a salesman and openly confess to you, respected Mr Nobel, that Sofie supports me to the present day. Indeed without her help I would have long been lost.

But you are indirectly our benefactor, respected Mr Nobel, and I am well aware of it since Sofie owns nothing, and I beg your pardon as Sofie's father, that she never told you. It was a noble motive that kept Sofie from doing so. She was concerned for us, feared that we might suffer and therefore preferred to suffer some hardship herself.

It was another coincidence that you, respected sir, finally discovered it, a happy coincidence for me, I must confess, since it finally presented me with an opportunity to approach you by letter, although I am taking

---

1 The southern part of Austria bordering Slovenia.
2 See Sofie's own complaints in SH 8.

the risk of offending you. But my spirit urges me on to tell you that Sofie has been very unhappy lately and had much trouble and problems with Olga,[3] I could tell you more in person and am convinced, respected Mr Nobel, that you would believe me.

Today I want to thank you sincerely on behalf of the family for all your good deeds, respected Mr Nobel, and cherish only one wish: to have the pleasure of meeting you in person, respected sir, and to tell you everything in conversation, how good and kind Sofie was all the time. Once more many thanks and greetings with singular respect,
Your devoted

H. Hess

Vienna, 25 May [1887]

Respected Sir,

I am in receipt of your respected letter and take the liberty – only *this once* – to respond in this matter as follows:

I acknowledge that in your arguments you have spared my feelings. I have learned several things now that I did not know before, and do not reproach you. You have acted and are acting honourably and like a gentleman and in a manner worthy of a man of culture and high intellect. Therefore I ask you not to give credence to every whispered suggestion, for Sofie has always spoken of you with the greatest love and respect. She could not have spoken and written of you in a nicer and more loving manner.

The accusation that Sofie's education was deficient concerns me as well, but I can assure you that Sofie was a good student, although she did not have a good memory and still hasn't. If Sofie had obeyed you then, it would no doubt have been to her advantage, but that cannot be changed now, and is secondary. Sofie's main task is to regain your trust by showing the correct understanding and appreciation for your honourable and respected name, and Sofie will also become more modest in her lifestyle, and generally avoid anything that is not right, I can

---

3 For Olga Böttger see Letter 33, and Sofie's catty remarks in SH 2, 5, 6.

guarantee you that today already, for my intention is always to keep before Sofie's eyes the enormous gratitude she owes to you.

Respected sir, you yourself say in your respected letter that you like Sofie's childlike nature and that she has not quite lost it as yet. In this sense you must judge this last unpleasant incident[4] and the aftermath, which was hurtful for both of you. Sofie was alone in the hotel here all through the winter and felt perhaps neglected by you and therefore very unhappy, and a lady alone in a hotel is exposed to all sorts of defamation, especially when they see that the lady is wearing nice clothes and generally lives well. For that reason Sofie was always on edge and often said to me: If only I hadn't come here! Forgive my kind Sofferl therefore if she had contact with Dr H.[5] I am sure it was not an intimate relationship. She may have felt encouraged to act in this manner because you told H. when you were in Merano that he should try to win Sofie's heart and that you would arrange for his marriage with Sofie, if she was agreeable of course. You also obliged H. in financial matters. Of course he still hasn't found it convenient to pay you back, and I therefore beg your pardon if I mention it. I only mean to say that it caused my daughter to believe that you were serious in that undertaking, but later she turned away from H. because you formed the correct judgment about H., which I endorse, and I thank you for showing such fatherly concern for Sofie in vetoing this connection. If H. nevertheless continued with his attentions to her with nice words and letters that completely lacked truth, his purpose was to obtain more money from you, but thank God he did not succeed. It is Dr H. who has caused much trouble and who is fully responsible for it. A woman is easily fooled. H. is culpable therefore.

You mentioned in your respected letter that I should not have written a reproachful letter to Mrs Böttger.[6] I didn't write a reproachful letter, I merely noted that Olga as she grew up formed her own opinion of the relationship between you and Sofie and therefore did not behave well and constantly angered Sofie, and I can attest to that.

It is not in my character to hurt anyone and I do not begrudge good things even to my worst enemy, but I despise all bragging. I hope you

---

4 See Letter 110a.
5 Hebentanz; see Letter 110a and note.
6 Olga's mother.

are not offended that I importune you with my respectful letter. I took the liberty to reply – only once, as I said.

I am very grateful that you will continue looking after my daughter to some extent and indulge in the fond hope that you will not give the impression to Sofie that she should not have helped us, for it is only under that condition that I will accept your assistance.

I wish you the best of health and send you sincere greetings.

Signed with special respect,

Your devoted H. Hess
II Kaiser Josefstrasse 5

Ischl, 28 July 1887

I was truly displeased when I received your respected letter of 24 July and felt very hurt by the accusations and defamations you listed against my daughter, which are based on the vengeful statements of Olga and may have been encouraged by her mother to harm my daughter again and to harm her health through hurtful remarks, in which she succeeded unfortunately, for Sofie is so nervous and rattled that she will be seriously ill if things go on as they do.

I myself have been ailing for some time, and your respected letter had the effect of making me take to my bed that very day. The following day, however, I decided to come here, even though I wrote to Sofie to come to Vienna immediately, but I did not want to raise any new suspicions, I told my daughter what had happened, a very sad mission for me. I did not mention your list of her misdeeds because you communicated them to me confidentially, I merely said to her "So what happened now?" since you asked me to guarantee that she will respect the conditions you set for her. In the meantime you sent a telegram addressed to Sofie Hess, "Villa must be vacated."[7] But think what scandal that would cause, if she was thrown out – bang, out of the blue – after an intimate relationship of ten years, and you would be the talk of the whole town yourself. Show some consideration for my family too, I flatter myself to have an honest name, have unmarried daughters, and if this comes out you ruin my family and make Vienna impossible to

---

7 I.e., the villa in Ischl. Nobel was planning to sell it (see Letter 175).

me. You will drive me to despair, and I therefore beg you to retract this ultimatum and let her stay until the end of August. I would not have allowed my younger daughter Anna to come here if I had had an inkling that there would be another rift between you and Sofie. Staying here in the circumstances isn't amusing, and certainly has no health benefits for Sofie, since she is constantly worried. My daughter lives a lonely life here, her only distraction is going for a daily drive and the fact that she has at least the company of my son and my daughter Anna. In your respected letter you mention an affair that supposedly happened eight years ago, I heard about that from my daughter then, and she had to suffer a great deal being under suspicion, and Sofie is prepared even today to justify herself in the presence of your nephew.[8]

Respected sir, you yourself say that hatred and revenge are outdated feelings for a man of reason, therefore you must not believe the defamations of little Olga, the truth is far from what she says, and don't believe that my daughter gave money to H.[9] She supported only me and her sisters and brother. In any case, Sofie does not maintain her relationship with you for the sake of money. She was pleased by your gentle manners and your generosity, which you always showed to her in the fullest measure, respected sir.

My daughter was never greedy for money. Sofie did not spend her youth with you to make money. She never thought of it, I asked her a few times whether she was sure that she was going to be looked after, but she always answered me with the words: "Alfred is such a sensitive man that I would not want to hurt his feelings by talking of money matters."

Concerning your letter I told her only that she hurt her interest a great deal, for you had designated a certain amount for her in your will, and she forfeited it. I must tell you frankly that it was all the same to her, whereas your cold manner now hurts her more than everything else.

---

8 See Letter 13 note. Two letters from Emmanuel Nobel to Sofie, dated 16 and 18 January 1879 respectively, shed light on the situation. The young man had spent an evening with Sofie. As he was about to leave she said, "Oh, it's raining outside. You must spend the night here." At least this is what he understood, but he conceded that he might have misunderstood Sofie. He furthermore attests that this was his only meeting with her and that nothing in their conversation that evening "was to her discredit." The letters are quoted (in translation) by Sjöman, *Mitt hjärtebarn*, p. 29. They were shown to him by Sofie's granddaughter, Olga Böhm.
9 See Letter 110a.

About the rest I would prefer to talk to you in person, and if that is not too unpleasant for you, would ask you to state a place where we can meet for this purpose.

I shall leave today and return [to Vienna], in case you want to send me a message to the address you have for me.

Accept my special respect and devotion, respected sir.

H. Hess

P.S. It would be a great pleasure to hear from my daughter that you will visit her so that she may feel a little better.

---

30 Oct 1887

Most respected Mr Nobel,

Yesterday late at night I received your telegram, and because it was addressed simply to "Hess," I opened it, but gave it to [Sofie] only this morning, together with the letter. I didn't want to disturb her night rest, since she told me that she sleeps poorly in any case. I am very sorry that you are not feeling well and hope you are feeling better when you receive my respectful letter of today.

I very much regret that I was not present during your last visit here. I had hopes of seeing you in Ischl since I was in the area on business and also read the telegram you sent to Sofie then: "Live with your parents or if need be at the Hotel Meissl until something suitable is found." Sofie understood this to mean, until you have found something suitable in Paris for her. She has been staying in the hotel for more than four weeks, and apart from the fact that it is expensive and also not very comfortable, people will talk even if there isn't the least reason. I visit Sofie daily in the morning and in the afternoon, Luis[10] comes at other times, as often as possible, and my daughters also visit Sofie from time to time.

Yet Sofie very much longs to get away from the hotel and would like best to stay with you. But if you do not want that, please tell her what she should or has to do finally to have a home of her own.[11] Sofie said to me the other day that Dr H[12] took care to slander her through a third party, and today you confirm this. Thus this gallant, sweet-talking

---

10 Sofie's brother.
11 The following year Nobel bought her a villa near Vienna. See Letter 129.
12 Hebentanz; see Letter 110a.

adventurer has changed his profession and has become a slanderer. Sofie speaks of him with disdain now and can call herself lucky that she didn't end up marrying him, for you saw quite through him and assessed him correctly, Mr Nobel.

The good thing is that you will have no reason in future to suspect that money goes to [Buda]pest, and what is more important, that Sofie has contact with him, when she has long given that up.

It would be a special pleasure to find that you have included a few lines to me in your letter, and I feel obliged to thank you for everything Sofie has done for me and billed to you. Accept my and my family's special respect and sincere greetings with the best wishes for your health. May you be happy until old age.

Your devoted H. Hess

B

LETTERS FROM NIKOLAUS KAPY

Reichenhall, 20 July 1895

Honoured Mr Nobel,

I beg you, for Heaven's sake don't desert us. Sofie and I are in Reichenhall completely without means because her illness has cost a great deal of money and day before yesterday the weekly hotel bill was due and had to be paid.

I regard you as a patron and benefactor of my future wife and therefore take the liberty to importune you.

Poor Sofie is so sick that it breaks my heart.

Our wedding will take place in September.

Begging you once again, I sign

Your devoted

Kapy

[P.S] Sofie knows nothing of this letter, she is staying here as "Mrs Kapy."

Vienna, 4 January 1896

Noble sir,

My wife submits to your wishes and has signed the agreement drawn up by your representative. At the signing Director Heidner[13] promised her to pay the debt accumulated up to the signing, ca. 2,000 gulden, and to speak on Sofie's behalf to you, noble sir, since those debts were contracted to cover various needs, especially medical costs. Furthermore, Director Heidner promised to redeem the jewellery my wife had to pawn. Since my wife will, and indeed must, live a modest and undemanding life from now on and will cover all her needs out of her present annuity of 500 gulden a month, and so that she will not burden you in future, it is my polite request that you, noble sir, will accept the wishes of my wife in that respect and give the appropriate instructions to Director Heidner.

   Asking you to fulfil my request, noble sir,

I remain your devoted Kapy

C

LETTERS FROM AMALIE BRUNNER

Celje, 8 April 1891

Dear Mr Nobel,

Already in January when Sofie visited us, she asked me to write to you and apologize to you on her behalf for her misstep.[14] I refused because it goes against the grain to appeal to your noble and good heart when she has committed such a great sin, a misstep that cannot be excused. Today Sofie wrote another pitiful letter, which made me truly feel her wretched and unhappy situation, and she asked me once more to write to you. I do it with a heavy heart and believe I can say only one thing in her defence: she is young and pretty, was left alone, abandoned in a big city, and in her loneliness could not withstand the temptations,

---

13 Sofie's trustee.
14 I.e., her affair with Kapy and the resulting pregnancy.

and yet she was loyal to you until then, or that misfortune would have befallen her sooner. I cherish the frail hope, dear Mr Nobel, that you in your well-known kindness will be a less severe judge of Sofie's misstep and forgive her. I have a great deal more to tell you about that, but that is easier said in person than put on paper. If you are in Italy this year in May or June, I beg you to give us joy and visit us.

Do not be angry that I importune you with these lines, dear Mr Nobel, you have a kind heart and will not desert the poor woman, who is very unhappy and made herself unhappy only because of her thoughtlessness. I console myself thinking that you will be kind and quietly hope that Sofie will become a better person after going through this sad schooling.

Sincere greetings from all of us, respected Mr Nobel.

I remain

Your devoted Amalie Brunner

Celje, 20 Feb 1892

Dear Mr Nobel,

I don't share the concern you express in your letter – that Sofie's debt is so large because she may be exposed to blackmailing. I am sure that the people to whom she gave receipts had no idea that she was not entitled to use that name. Her extravagance led her to borrow money at usurious rates, which of course made the debt grow horrendously. It would be best to redeem the receipts soon and to pay Dr Barber as well, even if he is Sofie's special friend.[15] Please don't mention that Sofie was not entitled to sign such bills, which would put him or some agents of his in a position to blackmail her. I can see that Sofie barely manages with her money because in the past she always sent me 5 gulden a month, to help me out a little, but for some time now she stopped doing that, although it is an amount that was nothing to her in the past.

Please be so kind to let me know to what extent you will take my advice concerning Sofie, since I find it rather strange that Sofie hasn't written to me in some time.

Sincere greetings,

Your devoted Amalie Brunner

---

15 *Hausfreund*, an ambiguous term that sometimes denotes an illicit sexual relationship.

Celje 13 April 1892

Dear Mr Nobel,

I was very happy to receive your friendly letter. It is most noble of you
once more to help my sister in her difficulties. I only wish she could
appreciate it and take it to heart. Don't you think it would be good to
talk about all that with Sofie in person? It is difficult to manage such
matters by letter. In your last letter you suggested that Sofie should
be put in the care of a trustee[16] or should marry – but whom? I believe
Sofie will be even more against marriage than she is against a trustee.
Something needs to be done, and in my modest opinion, the way out is
best discussed in person.

Sincere greetings from us all,

Your devoted Amalie Brunner

---

16 She was placed in the care of trustees in 1894. See her protests in SH 21.

Works Cited

Abbott, Elizabeth, *Mistresses: A History of the Other Woman*. London: Duckworth 2010.

Bergengren, Erik. *Alfred Nobel. The Man and His Work*. New York: T. Nelson, 1960.

Biedermann, Edelgard. *Der Briefwechsel zwischen Alfred Nobel und Bertha von Suttner*. Hildesheim: G.Olms, 2001.

Cvrcek, Tomas. "Wages, Prices, and Living Standards in the Habsburg Empire 1827–1911." *Journal of Economic History* 73, no. 1 (2013): 1–37.

Delvau, Alfred. *Les Plaisirs des Paris: Guide pratique et illustré*. Paris: A. Fauré, 1867.

Drumont, Edouard. *Le France Juive: Histoire Contemporaine*. Paris: C. Marpon & E. Flammarion, 1886.

Erlandsson, Ake. *Alfred Nobels bibliotek: En bibliografi*. Stockholm: Nobel Library of the Swedish Academy, 2002.

Fant, Kenne. *Alfred Nobel: A Biography*. Trans. by M. Ruuth. New York: Arcade, 2014.

Gaisbauer, Adolf. *Davidstern und Doppeladler: Zionismus und jüdischer Nationalismus in Österreich 1882–1918*. Cologne: H. Böhlaus, 1988.

Griffin, Victoria. *The Mistress: Histories, Myths, and Interpretations of the "Other Woman."* New York: Bloomsbury, 1999.

Hickman, Katie. *Courtesans: Money, Sex, and Fame in the 19th Century*. New York: Morrow, 2003.

Kienzl, Lisa. *Nation, Identität und Antisemitismus: Der deutschsprachige Raum der Donaumonarchie 1866–1914*. Graz: Vandenhoeck & Ruprecht, 2014.

Larsson, Ulf. *Alfred Nobel: Networks of Innovation*. Stockholm: Nobel Museum & Science History Publications, 2008.

Ley, Michael. *Abschied von Kakanien: Antisemitismus und Nationalismus im Wiener Fin de siècle.* Vienna: Sonderzahl Verlag, 2001.

Mackaman, Douglas P. "The Tactics of Retreat: Spa Vacations and Bourgeois Identity in Nineteenth-Century France." *Being Elsewhere: Tourism, Consumer Culture, and Identity in Modern Europe and North America,* ed. Shelley Baranowski and Ellen Furlough, 35–62. Ann Arbor: University of Michigan Press, 2004.

Mayer, Sigmund. *Ein jüdische Kaufmann 1831–1911.* Leipzig: Dunker & Humbolt, 1911.

Murray, John. *Handbook for Travellers to the Continent.* London: J. Murray, 1840.

Palmer, Francis. *Austro-Hungarian Life in Town and Country.* London: Newnes,1903.

Robertson, Priscilla. *An Experience of Women: Pattern and Change in Nineteenth-Century Europe.* Philadelphia: Temple University Press, 1982.

Robb, Graham. *Victor Hugo: A Biography.* New York: W.W. Norton & Co, 1997.

Ross, Andrew. *Urban Desires: Practicing Pleasure in the "City of Light" 1848–1900.* Doctoral diss., University of Michigan, 2011.

Sjöman, Vilgot, translator. *Mitt hjärtebarn: De länge hemlighållna breven mellan Alfred Nobel och hans älskarinna Sofie Hess.* Stockholm: Natur och Kultur, 1995.

Sohlmann, Ragnar. *The Legacy of Alfred Nobel: The Story behind the Nobel Prizes.* London: Bodley Head, 1983.

Schück, Henrik and Ragnar Sohlmann. *The Life of Alfred Nobel.* Uppsala, 1926; English translation London: Heinemann,1929.

Steward, Jill. "The Role of Inland Spas as Sites of Transnational Cultural Exchanges, 1750–1870." *Leisure Cultures in Urban Europe, c. 1700–1870: A Transnational Perspective,* ed. Peter Borsay and J.H. Furnée, 234–59. Manchester: Manchester University Press, 2015.

–. "The Spa Towns of the Austro-Hungarian Empire and the Growth of Tourist Culture: 1860–1914." *New Directions in Urban History: Aspects of European Art, Health, Tourism and Leisure since the Enlightenment,* ed. Peter Borsay, Gunther Hirschfelder, and Ruth-E. Mohrmann, 87–126. Münster: Waxmann, 2000.

Tolf, Robert W. *The Russian Rockefellers: The Saga of the Nobel Family and the Russian Oil Industry.* Stanford: Hoover Institution Press, 1976.

Vögtle, Fritz. *Alfred Nobel.* Hamburg: Rowohlt, 1983.

Wistrich, Robert. *The Jews of Vienna in the Age of Franz Joseph.* New York: Oxford University Press, 1989.

Index